SpringerBriefs in Energy

More information about this series at http://www.springer.com/series/8903

Yuping Huang · Panos M. Pardalos
Qipeng P. Zheng

Electrical Power Unit Commitment

Deterministic and Two-Stage Stochastic
Programming Models and Algorithms

 Springer

Yuping Huang
Department of Industrial Engineering
and Management Systems
University of Central Florida
Orlando, FL
USA

Qipeng P. Zheng
Department of Industrial Engineering
and Management Systems
University of Central Florida
Orlando, FL
USA

Panos M. Pardalos
Department of Industrial and Systems
Engineering
University of Florida
Gainesville, FL
USA

ISSN 2191-5520   ISSN 2191-5539   (electronic)
SpringerBriefs in Energy
ISBN 978-1-4939-6766-7   ISBN 978-1-4939-6768-1   (eBook)
DOI 10.1007/978-1-4939-6768-1

Library of Congress Control Number: 2016960772

Printed on acid-free paper

This Springer imprint is published by Springer Nature
The registered company is Springer Science+Business Media LLC
The registered company address is: 233 Spring Street, New York, NY 10013, U.S.A.

Preface

Electricity demand varies during each day and each week due to the cycling pattern of our life. In addition, electricity is an instantaneously perishable commodity and still cannot be efficiently stored in bulk. These facts raise an interesting question for electrical power generation: how to meet the time varying demands in the most economical way. To answer this question, a great amount of research efforts have been devoted to the Unit Commitment (UC) problem, which aims to optimally schedule the "on" and "off" statuses and power dispatches of electrical power generating units while considering multiple technical and economic constraints. The UC problems are mostly formulated as mixed integer linear programs. Based on different perspectives and purposes, there are many variants for the UC problem. These problems draw a lot of attentions from both power industry practitioners and academic researchers. Many types of algorithms have been developed for or applied to UC problems, such as dynamic programming, Lagrangian relaxation, general mixed integer programming algorithms, Benders decomposition, etc. This book focuses on two-stage stochastic unit commitment models and advanced techniques to efficiently solve the large-scale problems due to scenario propagation.

Orlando, USA   Yuping Huang
Gainesville, USA   Panos M. Pardalos
Orlando, USA   Qipeng P. Zheng
May 2016

Contents

Chapter 1
Introduction

Electric Power Systems

Electric power system is one of the most important service systems that keep our society running, as it is responsible for generating, transmitting and distributing electricity, which powers almost all aspects of our life. In United States, thousands of power systems are connected through electricity grids that are managed by Independent System Operators (ISOs) and Regional Transmission Organization (RTOs). As the three main components of electric power system, generation, transmission and distribution construct a multi-level network connecting initial energy supplies with end users for daily uses of electrical power.

Worldwide the major sources used to generate electrical power are still fossil fuels including coal, natural gas, oil, nuclear, etc., while renewable sources (e.g., wind, solar, etc.) are gaining bigger shares recently. According to the statistics from US Energy Information Administration (EIA) reports in 2014 [DOE14], existing power systems currently remain fossil-fuel dominated and this is linked to extensive Green House Gas (GHG) emissions and other pollutants. As shown in Fig. 1.1, the share of fossil fuel in total energy source still remains above 68% in 2012 and the renewable share of total energy sources (including biofuels) has grown up to 12%. Particularly, wind, solar thermal and photovoltaic energy have respectively 17% and 138% growth rates on the contribution to energy generation, compared to their historical data in 2011. We can expect a trend toward the mix of sources for net power generation shown in Fig. 1.2, where coal will still be dominant, followed by natural gas, nuclear and then renewable energy in the near future. Due to the implementations of effective energy policies and environmental policies, share of coal will continue to be reduced significantly while the renewable energy share of total generation source will increases up to at least 15% in 2025.

As electric power is primarily supplied by burning fossil fuels, coal-fired power plants has made the electric power sector the largest GHG contributor to global warming for a long time. The GHG emissions contain a vast majority of Carbon

© The Author(s) 2017
Y. Huang et al., *Electrical Power Unit Commitment*,
SpringerBriefs in Energy, DOI 10.1007/978-1-4939-6768-1_1

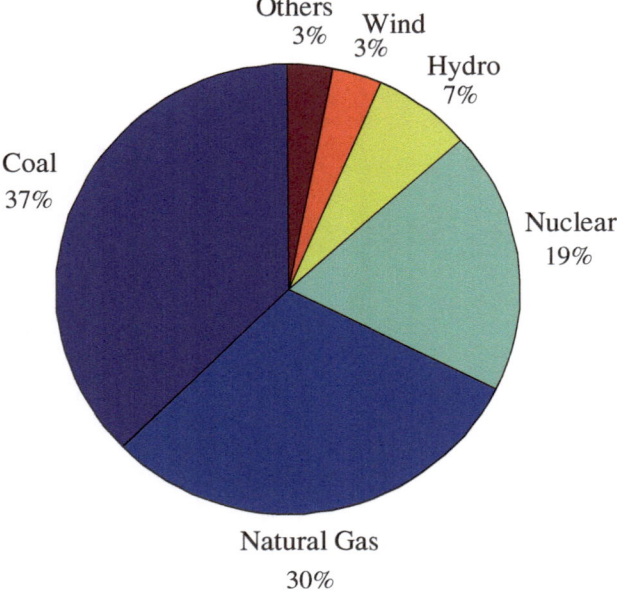

* *Others represents the sources from petroleum, other gas, solar, wood, geothermal, biomass and other energy sources.*

Fig. 1.1 Total electric power net generation, 2012 (Thousand Megawatt hours). Data source: [DOE14]

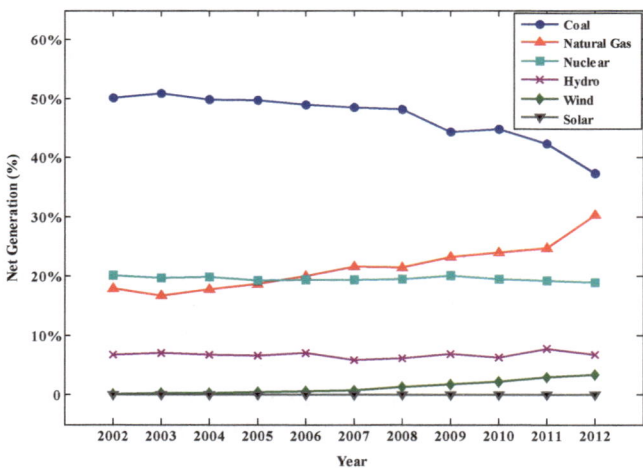

Fig. 1.2 Statistics for power net generation, 2002–2012 (Thousand Megawatt hours). Data source: [DOE14]

Dioxide (CO_2), a minority of methane (CH_4) and Sulfur Dioxide (SO_2), and lesser amounts of other gases. During 2012, the U.S. power industry produced 2,156,875 thousand metric tons of CO_2 which, although reduced by 11% of emissions compared to 2002, remain the largest source of GHG emissions. In response to mitigate climate change, the Environmental Protection Agency (EPA) has taken many actions to reduce GHG emissions from traditional coal-fired power plants including but not limited to increasing energy efficiency on power plants and end-use, adopting fuel switching, renewable energy portfolio as well as deploying carbon capture and storage (CCS) systems [HRZ13, ZRP+12, U.S14]. Among of them, carbon storage as the final step to prevent a large amount of CO_2 from being emitted to the atmosphere has been implemented successfully in geologic reservoirs and studied through optimization approaches for scheduling to sequestrate CO_2 while considering accompanying benefits [HZFA14, HRZ14].

The Impacts of Renewable Energy on Power Systems

More generally, renewable energy is defined as the energy offered through naturally and continually replenished resources, such as hydropower, wind power, solar power, biomass power, geothermal power among many others. Renewable energy is very attractive and sustainable because of "no" costs and/or no pollutant emissions, which well fits the current and future needs of the next-generation energy systems. However, due to the intermittent and uncertain nature of renewable energy, people realize that a fast-growing penetration of renewable energy to current power grids would bring a lot of challenges to the effective and efficient operations and management of power systems.

Currently, as the portion of renewable energy grows fast, power systems are required to become more flexible to accommodate the variability and uncertainty of renewable energy outputs. For example, in the cases with a high penetration of wind power, it's really important to predict as accurate as possible the wind energy output based on wind speed pattern and historical data. The deviations from forecasted wind outputs due to dramatic increase/decrease on wind speed would force conventional thermal generators to ramp up/down quickly to maintain the power balance. As this situation occurs frequently, the increasing variability and unpredictability of renewables generation systems would result in further intensifying generator cycling and increasing additional operational costs.

In addition to continuous uncertainties caused by renewable energy and demands, an unplanned outage of generators or transmission elements is considered a low-probability event and could occur in much low frequency. This type of unexpected uncertainty like power blackout can be covered by contingency control planning and can be treated through a robust optimization approach [SOA11, XJ12]. In most instances, power system is not built for avoiding any uncertainties, but an operational schedule for power generation should be a robust solution to handle the impacts from most of uncertainties.

The Generation Scheduling on Power Systems

Throughout power generation systems in practice, there is only a minority of power plants operating in isolation from power grids, while most of power plants participate in energy market and connect their resources to power grids. Based on available energy resources and forecasted loads, ISOs perform the main functionality of scheduling the operations of generation units and determining hourly market clearing prices for power market, and also perform energy procurement and congestion management in the real-time market. The power generation scheduling, also named unit commitment (UC), is essential for the whole power system operations from day-ahead operational schedules to real-time economic dispatch, even extended to contingency management.

Unit commitment is one of the classic optimization problems in power systems operations and control. Generally, unit commitment problems have two common objectives. One is used by ISOs to minimize the total operational cost mainly from thermal generators to meet a generation target or an forecasted hourly load, and the other is to maximize the total profit when GENCOs (GENeration COmpanies) make bidding strategies. Most of UC problems are formulated based on dynamic programming or mixed integer linear programming (MILP) methods. A UC model basically includes both binary decision variables to indicate the on/off status of generation units and continuous variables to indicate the dispatches and reserves, as well as power generation and operation constraints, such as capacity limits, minimum on/off hours and ramping constraints. Also, unit commitment has been developed to provide all kinds of operation scheduling solutions for balancing energy supply and demand in day-ahead and hour-ahead markets, which is implemented by ISOs in deregulated electricity markets (Fig. 1.3).

The classic unit commitment problem is the security-constrained unit commitment (SCUC) widely studied to minimize the total cost while maintaining a system's reliability at an expected level [SYL02]. Many researcher have proposed to adopt engineering techniques and constraints to address and solve the reliability issues, such as transmission constraints [FSL05b, FSL06], "n-1" criteria [HFO+10], stochastic demands [WSL08], etc. As there could be thousands of generation units and transmission lines existing in a system, the unit commitment problem modeled by MILP becomes a computationally challenging problem with a large number of integer variables and constraints. Various optimization techniques including Lagrangian relaxation and branch-and-bound based MILP methods have been used to solve large-scale UC problems [HROC01, SYL02]. Benders' Decomposition and Lagrangian Relaxation techniques have also been developed for specific UC problems to reduce computational expenses in the means of separating the master UC problem (determining the on/off statuses) from the reliability checking subproblems to make the original problem smaller and solvable [FSL05b, FSL06]. Meanwhile, the solution process will generate one or more Benders cuts from reliability constraints or contingency simulation subproblems, and then add those cuts to the master UC problem in the subsequent computational process when a requirement is violated [CCMGB06, FSL06].

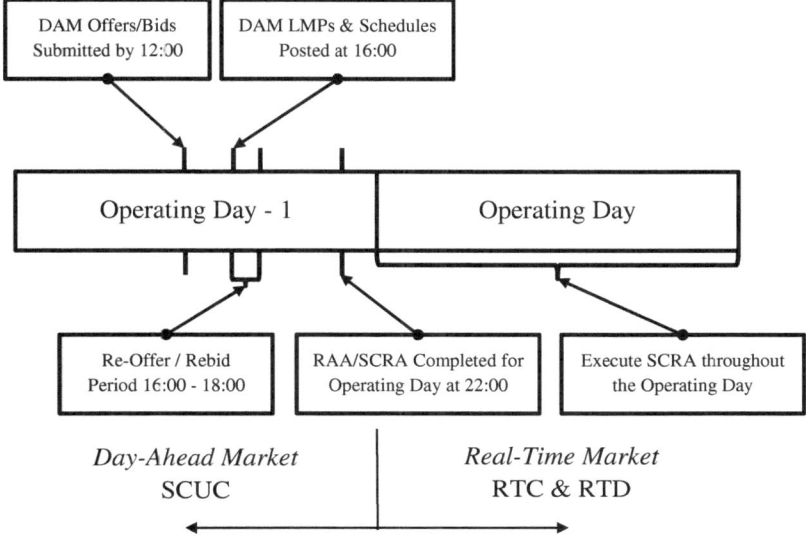

- DAM: Day-Ahead Market
- RTM: Real-Time Market
- RTC: Real-Time Commitment
- RTD: Real-Time Dispatch
- LMP: Locational Marginal Price
- RAA: Reserve Adequacy Assessment
- SCRA: Security-Constrained Reliability Assessment
- SCUC: Security-Constrained Unit Commitment

Fig. 1.3 Day-Ahead market and real-time market timeline [Joh10, ISO16]

Unit Commitment under Uncertainty

As future energy needs keep growing, the conventional unit commitment shows a lot of restrictions to keep up with increasing changes. ISO would need to implement innovative changes on energy and ancillary markets to accommodate the supply-and-demand challenges. For achieving higher reliability of power systems, ISOs plan to implement the market process and scheduling improvements using state-of-the-art unit commitment models that are fully based on operations management and optimization methods. Also, ISOs expect to better integrate renewable energy into existing systems, reducing the effects of the intermittence and variability of supplies and demands on already scheduled commitments of units. One can seek an applicable solution from three following perspectives to deal with uncertainties in power systems, including,

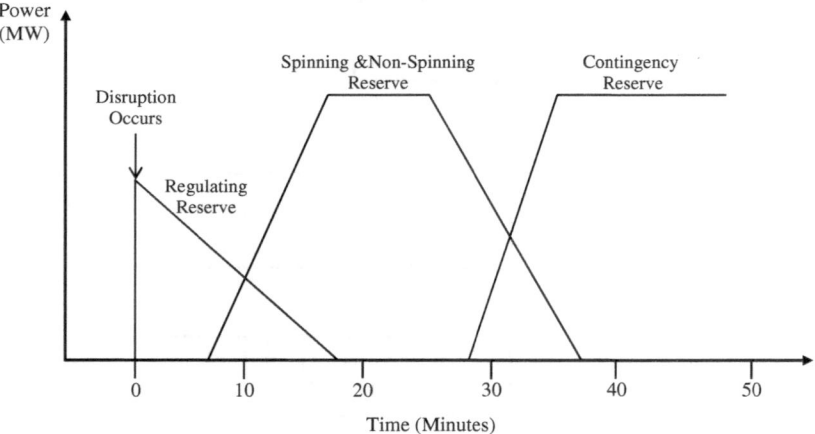

Fig. 1.4 The general timeline of operating reserve

- implementing reserve requirements and providing related reserve services,
- adopting non-generation resources, and
- applying advanced solution methods, such as stochastic optimization and robust optimization.

First, operating reserve is a widely used approach in the power industry to deal with uncertainties on power systems. Generally, a part of generating resources will be retained in order to handle unexpected surges or contingency events. The current operating reserve is comprised of spinning reserve and non-spinning reserve, in addition to regulating reserve and contingency reserve. They have different main functionality and can be provided or procured according to generator's characteristics or commitment from different energy sources. And, the operation timeline for each reserve will be executed after an unexpected disruption occurs, shown in Fig. 1.4. The regulating reserve is to provide the automatic response on output frequency mainly consisting of regulation up and regulation down, and it is followed by spinning and non-spinning reserves that can cover power shortage within up to 30 min, while the contingency reserve is the last backup with the goal of restoring operating reserve and may overlap with non-spinning reserve. Since above reserves don't require specific new technology or operating requirements, they have been successfully implemented in generating operations to mitigate uncertainties for a long time.

Next, non-generation resource is viewed as a non-conventional source of energy and has been proposed to diversify power market services and further improve the stability, flexibility and reliability of energy supply. In the view of ISOs, non-generation resources consist of demand response, energy storage and other non-generation dispatchable resources to support the power balance. Through a decade of research, non-generation techniques and programs, particularly in demand response and energy storage, have been well studied and developed for expanding use of renewable energy as well as helping improve the cost-effectiveness of renewable energy systems. People have know their advances and benefits for power generation, transmission, and end users, but some technical and operational issues are still waiting for solutions before they can be incorporated with existing power systems and implemented in a wide range. As the redesign of ISO wholesale market still a work in progress, CAISO has attempted to allow non-generation resources to enter the ISO regulation markets and be used for regulation services [Cal10, Ang12].

What's more, taking the advantages of operations research, its applications in energy systems can aid the better integration of existing power plants with renewable energy sources at the strategic level. At the operation level, these methods and solution techniques are also beneficial for power system operation scheduling in practice and making full use of energy from accessible resources while meeting modern reliability needs simultaneously. Recently, more advanced modeling techniques methods are used to address the variability and uncertainty brought by uncertain demands and renewable energy sources, among which stochastic unit commitment (SUC) has emerged as one of the most promising tools [BBMW06, RPZ+09, TMDO09, WBC+09].

Different from traditional UC problem, the stochastic unit commitment problem capture the uncertainty and variability of the underlying factors by simulating a certain number of scenarios [ZWL15]. Each scenario is expressed as a possible realization of the uncertain sources, e.g., wind output, demand, or fuel price. One can simulate many scenarios to represent an uncertainty to a large extent. However, we would face a series of computational challenges because the large number of scenarios dramatically increases computational complexity. Thus, more advanced optimization solution techniques are proposed to solve the large-scale system problems a within reasonable time. The computational improvement also allows optimization methods and techniques become the powerful tools for the optimization of power generation scheduling, but also open another opportunity to improve operational performance in the future research.

Structure of this Book

When ISOs perform the generation scheduling for energy markets (day-ahead, hour-ahead and real-time), solving short-term unit commitment problems will need to have full consideration of all different requirements, involving renewable energy generation, demand variation, operating and spinning reserves, transmission planning, non-generation resources, risk aversion, contingency, unit maintenance, and so on. Nevertheless, the aggregated requirement bring a lot of challenges on dispatch planning, solution implementation, resource efficiency, and system reliability as well. Decision makers may have to achieve serval organizational goals or specific decision requirements, such as seeking a most conservative decision to meet all demands based on the worst instance, or looking for an aggressive solution to reduce operational costs but maximize the use of renewable energy when the real adjusted price (with environmental and social aspects incorporate) keeps shooting up. However, the scheduling process requires operations research tools and methods to understand and ensure all decisions to be able to accomplish generation, transmission and distribution tasks with the objectives of low operation costs, high resource utilization, high flexibility and reliability to handle unexpected supply/demand fluctuations.

The main goal of this book is to introduce the recent development of electrical power unit commitment within the context of optimization modeling and algorithms. The studies in the last decade provide many useful ways of gaining new insights into the operations management of power systems. Without supply and demand uncertainties, deterministic unit commitment problems have been well addressed and help to build a solid foundation for solving more complicated unit commitment problems in the presence of uncertainty. Also, the ongoing research of two-stage stochastic unit commitment demonstrates that the development of power system reliability can be achieved by using advanced optimization methods. The main advantage from this research lies in the ability of handling predictable uncertainty, particularly that it can help integrate and apply renewable energy and non-generation resources into the daily power system operations.

This book thus consists of two main chapters to discuss unit commitment problems plus appendices:

- Deterministic Unit Commitment Models and Algorithms (Chap. 2).
- Two-Stage Stochastic Programming Models and Algorithms (Chap. 3).
- Appendices on common terms, model nomenclature, and a method to generate renewable energy scenarios.

Chapter 2 focus on basic unit commitment modeling based on mixed integer linear programming (MILP) and corresponding solution approaches. The major theme of this chapter is to discuss what objective an operation scheduling should achieve, what requirements a power system operation must satisfy as well as common solution techniques applied to solve MILP models. Chapter 3 mainly covers the advanced studies carried out for modeling the unit commitment problems associated with the participation of non-generation resources, ancillary service, contingency, real-time rescheduling and risk management through stochastic optimization approaches. The advanced concepts and mathematical modeling in SUC problems are introduced in details and also case studies are given correspondingly by using enhanced computational algorithms. To satisfy higher requirements on security, economic, and system reliability, power system operations will continue be incorporated with advanced operations research methods and applications to mitigate impacts arising from uncertainties and new technologies and societal developments.

References

[Ang12]   Angelidis G (2012) Integrated day-ahead market. Technical description, California ISO
[BBMW06]   Barth R, Brand H, Meibom P, Weber C (2006) A stochastic unit-commitment model for the evaluation of the impacts of integration of large amounts of intermittent wind power. In: International conference on probabilistic methods applied to power systems, pp 1–8, June 2006

[Cal10]   California independent system operator. Draft final proposal for participation of non-generator resources in California ISO ancillary service markets. Technical report, California ISO (2010)

[CCMGB06]   Conejo AJ, Castillo E, Mínguez R, García-Bertrand R (2006) Decomposition techniques in mathematical programming - engineering and science applications. Springer, Heidelberg

[DOE14]   DOE/EIA. Annual energy outlook 2014 with projections to 2040. Technical report, U.S. Energy Information Administration (2014)

[FSL05b]   Fu Y, Shahidehpour M, Li Z (2005) Security-constrained unit commitment with AC constraints. IEEE Trans Power Syst 20(2):1001–1013

[FSL06]   Fu Y, Shahidehpour M, Li Z (2006) AC contingency dispatch based on security-constrained unit commitment. IEEE Trans Power Syst 21(2):897–908

[HFO+10]   Hedman KW, Ferris MC, O'Neill RP, Fisher EB, Oren SS (2010) Co-optimization of generation unit commitment and transmission switching with N-1 reliability. IEEE Trans Power Syst 25(2):1052–1063

[HROC01]   Hobbs BF, Rothkopf MH, O'Neil RP, Chao H (2001) The next generation of electric power unit commitment models. Kluwer Academic Publishers, Norwell

[HRZ13]   Huang Y, Rebennack S, Zheng QP (2013) Techno-economic analysis and optimization models for carbon capture and storage: a survey. Energy Syst 4(4):315–353

[HRZ14]   Huang Y, Rahil A, Zheng QP (2014) A quasi exact solution approach for scheduling enhanced coal bed methane production through CO_2 injection. Optimization in science and engineering. Springer, New York, pp 247–261

[HZFA14]   Huang Y, Zheng QP, Fan N, Aminian K (2014) Optimal scheduling for enhanced coal bed methane production through CO_2 injection. Appl Energy 113:1475–1483

[ISO16]   ISO New England Inc. ISO New England manual for market operations manual M-11. http://www.iso-ne.com/participate/rules-procedures/manuals. Accessed May 2016

[Joh10]   Johnson S (2010) NYISO day-ahead market overview. In: FERC technical conference on unit commitment software, pp. 1–29

[RPZ+09]   Ruiz P, Philbrick C, Zak E, Cheung K, Sauer P (2009) Uncertainty management in the unit commitment problem. IEEE Trans Power Syst 24(2):642–651 May

[SOA11]   Street A, Oliveira F, Arroyo J (2011) Contingency-constrained unit commitment with $N - k$ security criterion: a robust optimization approach. IEEE Trans Power Syst 26(3):1581–1590

[SYL02]   Shahidehpour M, Yamin H, Li Z (2002) Market operations in electric power systems. Wiley, New York

[TMDO09]   Tuohy A, Meibom P, Denny E, O'Malley M (2009) Unit commitment for systems with significant wind penetration. IEEE Trans Power Syst 24(2):592–601

[U.S14]   U.S. EPA. Sources of greenhouse gas emissions: Electricity. http://www.epa.gov/climatechange/ghgemissions/sources/electricity.html. Accessed July 2014

[WBC+09]   Wang J, Botterud A, Conzelmann G, Miranda V, Monteiro C, Sheble, G (2009) Impact of wind power forecasting on unit commitment and dispatch. In: 8th International wind integration workshop, Bremen, Germany, October 2009

[WSL08]   Wang J, Shahidehpour M, Li Z (2008) Security-constrained unit commitment with volatile wind power generation. IEEE Trans Power Syst 23(3):1319–1327

[XJ12]   Xiong P, Jirutitijaroen P (2012) An adjustable robust optimization approach for unit commitment under outage contingencies. In: 2012 IEEE power and energy society general meeting, pp 1–8

[ZRP+12]   Zheng QP, Rebennack S, Pardalos PM, Pereira MVF, Iliadis NA (2012) Handbook of CO_2 in power systems., Energy systemsSpringer, Heidelberg

[ZWL15]   Zheng QP, Wang J, Liu AL (2015) Stochastic optimization for unit commitment-a review. IEEE Trans Power Syst 30:1913–1924

Chapter 2
Deterministic Unit Commitment Models and Algorithms

This chapter introduces the basic formulations of unit commitment problems which are generally proposed to optimize the system operations by mixed integer linear programming. Meanwhile, the formulations target a series of external factors that affect electrical power generation schedules, such as ramping capacity, reserve requirement, transmission capacity, fuel constraint and emission. This chapter also introduces the solution approaches to solve the deterministic unit commitment problems, especially using Lagrangian Relaxation and Benders' Decomposition. The SCUC cases are provided to illustrate the UC modeling and decomposition processes. All formulation notations are listed in Appendix B for reference.

2.1 Introduction

Generally, unit commitment is defined to optimize the ON/OFF status of generating units to meet the forecasted loads and reserve requirements, so as to provide a least-cost power generation schedule. The unit commitment problems namely consider how to optimally operate generators under physical conditions, such as generation capacity, minimum ON time, minimum OFF time, ramp up/down rate, reserve requirements, as well as generation costs, such as startup/shutdown cost and fuel costs.

Since the electric power generation is not an isolated component in the power system, the real-time dispatch levels are also subject to demand changes, transmission capacity and corresponding transmission conditions. Assuming that the real-time loads follow the expectations of forecasted loads, when the transmission outage possibly occurs at a time, it would cause to transmission congestions in some lines and change the original transmission flows on current networks, and meanwhile, will likely affect the original power generation schedule (real-time unit commitment). This correlation reveals the importance of the co-optimization of generation and transmission in practice. Although the unit commitment problems combined with a transmission constrained network become more complicated, these studies are very helpful to guide unit commitment scheduling from the perspective of a whole power system's operations.

© The Author(s) 2017   11
Y. Huang et al., *Electrical Power Unit Commitment*,
SpringerBriefs in Energy, DOI 10.1007/978-1-4939-6768-1_2

2.2 Objective Function

The objective function of unit commitment usually achieve the minimum total operational cost over a planned time horizon, the maximum social welfare or the maximum total profit for a GENCO.

A generic UC objective function is composed of two component costs, related to two-stage decisions. The first component cost is determined by day-ahead decisions, i.e. the startup decision and shutdown decision on each generator (in first stage). We here assume there will be no reschedule of units occurring during next-day operating hours. The first-stage decision includes the start-up decision v_{gt} and the shutdown decision w_{gt} that indicate when generation units will be turned on or shut down, and other operational determinations for operation services. The second component cost comes from the total operational costs in the second stage, which is primarily made up of fuel cost and possible unserved energy penalty. And, this unserved energy penalty is usually produced by load-shedding losses when scheduled generators are not able to satisfy real-time demands. There is a list of parameter definitions in Table 2.2 for reference.

$$\min \ \sum_{g \in G} \sum_{t \in T} (SU_g v_{gt} + SD_g w_{gt}) + \sum_{g \in G} \sum_{t \in T} F_g(p_{gt}) + VOLL \sum_{i \in N} \sum_{t \in T} \delta_{it}$$

$$(2.1)$$

where

SU_g	start up cost of unit g
SD_g	shut down cost of unit g
$F_g(\cdot)$	fuel cost function for unit g
p_{gt}	the thermal power generation/dispatch amount of unit g at time t
$VOLL$	value of loss load [\$/MWh]
δ_{it}	load loss at but i at time t

It should be noted that the fuel cost in the second-stage objective function is a quadratic function highly associated with power dispatch on a generator and fuel price. In general, the fuel cost function can be presented as a quadratic function of the dispatch/production level, p, i.e., for a generator g, $F_g(p) = a + bp + cp^2$, where a, b and c are usually positive cost coefficients. We know that the quadratic mixed 0–1 integer programming problem is not easy to solve in practice, especially when a lot of generators are involved. Further, due to the presence of binary decisions, this could bring extra computational burden on solving a nonlinear fuel cost function.

Instead of solving the mixed integer quadratic problem, an alternative method is to apply the piecewise linear approximation method to gain very close solutions for computational convenience. In other words, the original objective function is reformulated to generate a piecewise linear approximation and become a mixed integer linear programming problem. For gaining the piecewise linear approximation, the sum of squares (SOS) techniques are often used to substitute the fuel cost function $F_g(p)$ by the summation $\sum_{k=1}^{K} C_k \lambda_k$ with additional constraints,

$$\{p_g = \sum_{k=1}^{K} \Delta_k \lambda_k, \sum_{k=1}^{K} \lambda_k = u_g, \lambda_k \geq 0, k = 1, \ldots, K\},$$

where u is the commitment status of generator g, and C_k and Δ_k are coefficients used to approximate the quadratic curve.

Based on two status of a generator, we can know that when a generator is online and commits to supply capacity, i.e. $u_g = 1$, the UC model will be introduced with the following constraints,

$$p_g = \sum_{k=1}^{K} \Delta_k \lambda_k,$$

$$\sum_{k=1}^{K} \lambda_k = 1,$$

$$\lambda_k \geq 0, \quad k = 1, \ldots, K.$$

When the generator commitment status is in an "off" state, i.e., $u_g = 0$, the power dispatch level p_g become zero and has no any operational cost $F_g(p)$.

Because the cost function itself is convex (see Fig. 2.1), the piecewise linear approximation function is still convex. The solution obtained from the MILP is very close to the real optimal solution [HROC01, ZWPG13].

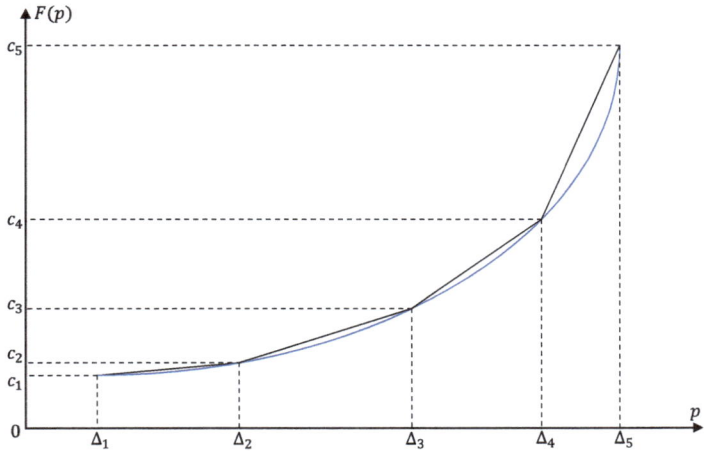

Fig. 2.1 Piecewise linear approximation of the fuel cost function [ZWPG13]

2.3   Constraints

In this sections, we introduce several common sets of UC constraints and variables from two-stage mixed integer linear programming models in details. From the most recent studies, we separate those typical constraints to address based on operation characteristics and service requirements.

2.3.1   Unit Commitment Constraints

In the day-ahead markets, an ISO determines an unit commitment schedule based on forecast demands and bids before the operating day, and designates power plants to prepare to generate electricity for next-day demands. As the first stage of operation scheduling, the UC constraints state generator status restricted by specific operation requirements, such as minimum ON time and minimum OFF time, and also specify startup action and shutdown action on each unit at a time period t, respectively.

Because a generator can't be started up or shut down arbitrarily in consecutive hours, Constraints (2.2) and (2.3) respectively indicate two generator's requirements: the shortest ON duration has to be met before a generator being shut down and the shortest OFF duration is also required before a generator being restarted up.

minimum ON time constraint:

$$u_{gt} - u_{g(t-1)} \leq u_{g\tau} \qquad \forall\, g \in G,\ t \in T, \tau = t, \dots, min\{t + L_g - 1, |T|\} \quad (2.2)$$

minimum OFF time constraint:

$$u_{g(t-1)} - u_{gt} \leq 1 - u_{g\tau} \qquad \forall\, g \in G,\ t \in T, \tau = t, \dots, min\{t + l_g - 1, |T|\} \quad (2.3)$$

where

u_{gt}:   commitment decision, a generator commits online, if $u_{gt} = 1$; otherwise, $u_{gt} = 0$.
L_g: the minimum-ON duration
l_g: the minimum-OFF duration
τ: time alias, a possible operating time period starting from time t
$|T|$:   the duration of a planning horizon

The startup action v_{gt} and the shutdown action w_{gt} are determined by the generator commitment statuses in the previous time period $t - 1$ and the current time period t. Any operational actions can incur startup or shutdown costs, which are considered in the objective function.

Startup action constraint:

$$v_{gt} \geq u_{gt} - u_{g(t-1)} \qquad\qquad \forall\, g \in G,\, t \in T \qquad (2.4)$$

Shutdown action constraint:

$$w_{gt} \geq -u_{gt} + u_{g(t-1)} \qquad\qquad \forall\, g \in G,\, t \in T \qquad (2.5)$$

$$u_{gt},\, v_{gt},\, w_{gt} \in \{0, 1\} \qquad\qquad \forall\, g \in G,\, t \in T \qquad (2.6)$$

where

v_{gt}: binary variable, startup action of unit g at time t
w_{gt}: binary variable, shutdown action of unit g at time t

2.3.2 Thermal Generation Constraints

According to given unit commitment schedules, power generation is to fulfill system operations through available generation resources and then to provide the least-cost generation outputs to serve demand. A generator output in a hour is subject to the maximum generation limit P_g^{max} and the minimum generation limit P_g^{min}. When a generator is scheduled online ($u_{gt} = 1$), the generation capacity is active giving bounds on dispatch level, shown in (2.7); otherwise, a generator output is forced to zero.

$$P_g^{min} u_{gt} \leq p_{gt} \leq P_g^{max} u_{gt} \qquad\qquad \forall\, g \in G,\, t \in T \qquad (2.7)$$

$$p_{gt} \geq 0 \qquad\qquad \forall\, g \in G,\, t \in T \qquad (2.8)$$

In addition, a generator output can be adjusted, increasing or decreasing between two successive time periods. The generation difference between two adjacent time periods is called ramping. A basic constraint to address generation ramping is presented in (2.9).

$$-RD_g \leq p_{gt} - p_{gt-1} \leq RU_g \qquad\qquad \forall\, g \in G,\, t \in T \qquad (2.9)$$

where

RD_g: ramp-down rate of unit g
RU_g: ramp-up rate of unit g

We also take into account some specific ramping situations, in which ramping rate is a changeable value and affected by the previous time period of commitment status. T Some recent models have addressed this situation [WWG13c, WWG13b]. If a generator has a startup ramping, i.e. the dispatch level ramping up from 0 MW to P_g^{min}, the regular ramp up rate is not suitable under this condition, but can be replaced with P_g^{min}. In addition to startup ramping or shutdown ramping, the regular ramp up rate and the ramp down rate are applied to consecutive online status. Therefore, constraint (2.9) can be modified as follow:

$$p_{gt} - p_{gt-1} \leq P_g^{min}(2 - u_{gt} - u_{g(t-1)}) + RU_g(1 + u_{g(t-1)} - u_{gt}) \quad \forall\, g \in G, t \in T \tag{2.10}$$

$$p_{gt-1} - p_{gt} \leq P_g^{min}(2 - u_{gt} - u_{g(t-1)}) + RD_g(1 - u_{g(t-1)} + u_{gt}) \quad \forall\, g \in G, t \in T \tag{2.11}$$

where constraint (2.10) describes two following situations:

- If a unit is ON at time $t - 1$ and ON at time t, the ramp up rate is RU_g;
- If a unit is OFF at time $t - 1$ and ON at time t, the ramp up rate is P_g^{min}.

Similarly, constraint (2.11) describes other two situations:

- If a unit is ON at time $t - 1$ and ON at time t, the ramp down rate is RD_g;
- If a unit is ON at time $t - 1$ and OF at time t, the ramp down rate is P_g^{min}.

2.3.3 Operating Reserve Constraints

Operating reserve is one type of ancillary operations to support the power balance on the demand sides. The ISO promote ancillary services not only to enlarge the pool of energy resources and introduce advanced techniques that effectively and actively participate in the ISO market, but also to support the renewable energy integration as a complementary tool.

The current operating reserve services being offered in electric energy markets include synchronous or non-synchronous, regulation reserves, spinning reserves, and non-spinning reserves. The sources of energy provided from different reserve services are different: regulation service mainly supplied from online generators, partial spinning reserve provided from generators already connected to the grid or system resources, and non-spinning reserve provided from quick-start generators, system resources or interruptible loads. The response times of reserve services cn vary from a few seconds to 30 min, up to 60 min, depending on the control reserve deployment time.

To achieve the optimization of energy and reserve in practice, one can obtain an efficient energy and reserve offering strategy by Heuristic method [NLR04] or consider the reserve determination on pre-contingency and post-contingency conditions [BGC05]. In the fact that the durations of reserve services are often less than 30 min, if the reserve duration is considered as a significant factor, a sub-hourly unit commitment model become necessary to handle this time transition issue [YWGZ12]. Here, we primarily focus on hourly unit commitment formulations based on an optimization method.

The spinning reserve is generally accounted for partial online generating capacity or off-line generation resources. Their outputs are constrained by predetermined maximum spin reserve, shown as

$$0 \leq s_{gt} \leq S_g^{max} \qquad\qquad \forall\, g \in G, t \in T. \qquad (2.12)$$

where

s_{gt}: spinning reserve of unit g at time t
S_g^{max}: maximum spinning reserve limit of unit g

Meanwhile, the generators that participate in biding spinning reserve must meet the spin reserve requirements given by ISOs. Constraint (2.13) describes an operating condition that the total spinning reserve at bus i should not less than the fixed reserve requirement.

$$\sum_{g \in G_i} s_{gt} \geq RS_{it} \qquad\qquad \forall\, i \in N,\, t \in T \qquad (2.13)$$

where

RS_{it}: spinning reserve requirement for bus i at time t.

More typical constraints regarding spinning and non-spinning reserve requirements are shown in constraints (2.14)–(2.18). The provisions of spinning reserve are expended, not only from internal spinning reserves (e.g. from synchronized generators) but also from external spinning reserve (purchased from spinning reserve not served). The maximum spinning reserve can be estimated through the response time of spinning reserve at ON status, which is shown in (2.15).

$$\sum_{g \in G_i} s_{gt} + (sn)_t \geq RS_{it} \qquad\qquad \forall\, i \in N,\, t \in T \qquad (2.14)$$

$$0 \leq s_{gt} \leq SRT \times MSR_g \times u_{gt} \qquad\qquad \forall\, g \in G,\, t \in T \qquad (2.15)$$

The non-spinning reserve has a more complicated situation, in fact, divided into two types of reserves: nonspinning reserve if a unit is ON and nonspinning reserve if a unit is OFF. The former nonspinning reserve is similar to regular spin reserve from online generators, and the latter nonspinning reserve is provided from off-line quick start generators with a higher level of nonspinning capacities. Either of nonspinning reserve is also necessary to satisfy the non-spin reserve requirements by the total provisions of non-spin reserve resources. The corresponding formulations are given in (2.16)–(2.18).

$$\sum_{g \in G_i} \left(ns_{gt}^{ON} + (ns)_{gt}^{OFF} \right) + (nsn)_t \geq NRS_t \qquad\qquad \forall\, i \in N,\, t \in T \qquad (2.16)$$

$$0 \leq (ns)_{gt}^{ON} \leq NSRT \times MSR_g \times u_{gt} \qquad\qquad \forall\, g \in G,\, t \in T \qquad (2.17)$$

$$0 \leq (ns)_{gt}^{OFF} \leq QSC_g (1 - u_{gt}) \qquad\qquad \forall\, g \in G,\, t \in T \qquad (2.18)$$

All reserves mentioned above are dominant in ancillary service markets. Meanwhile, more and more new products like flexible ramping products will be added to ancillary services and enrich the ancillary services market. ISOs also expect to benefit from the co-optimization by the effective determination of market clearing prices, the enhancement of reserve shortage pricing, the identification of units for system re-dispatch and proper compensation, etc.

2.3.4 Transmission Constraints

Power flows in a transmission network are usually considered in UC optimization problems, because they can be used to address power losses occurring in a network and eventually affect real-time power dispatch at a bus. Generally, Kirchhoff's current and voltage laws in a nodal way are applicable to find out electricity characteristics of transmission and distribution systems. Through simplifying calculation processes, one can present the power transmission using a DC linear approximation of power flows. In addition to voltage magnitudes, MVA or MVAR flows, the DC power flow method actually is often used to determine the MW flows on transmission lines in optimization models.

Measuring load-shedding losses is to help decision makers identify possible load losses at a specific bus. We can introduce a loss variable δ_{it} into the DC approximation of KCL constraints, in which the loss appeared at a bus for each time period will cause unserved energy penalty. The modified DC approximation of KCL involves in-bound and out-bound flow, thermal generation, forecasted demands, renewable energy generation as well as load-shedding loss, shown in constraint (2.19). The power transmission line from bus i to j also has a flow limit given in (2.20). In some cases, the load-shedding loss is not allowed in a specific location and thus δ_{it} needs to be restricted to zero.

$$\sum_{(i,j)\in A_i^+} f_{ijt} - \sum_{(j,i)\in A_i^-} f_{jit} = \sum_{g\in G_i} p_{gt} + R_{it} - D_{it}^0 + \delta_{it} \quad \forall\, i \in N,\, t \in T \quad (2.19)$$

$$- F_{ij}^{max} \leq f_{ijt} \leq F_{ij}^{max}, \qquad\qquad\qquad \forall\, (i,j) \in A,\, t \in T \quad (2.20)$$

$$l_{it} \geq 0, \qquad\qquad\qquad\qquad\qquad \forall\, i \in N,\, t \in T \quad (2.21)$$

where

f_{ijt}:	unrestricted variable, a bi-direction flow between bus i and bus j
δ_{it}:	load-shedding loss at bus i at time t
A_i^+:	the set of flow starting at bus i
A_i^-:	the set of flow ending at bus i
R_{it}:	renewable energy output at bus i at time t
F_{ij}^{max}:	transmission flow limit between bus i and bus j

Additionally, a DC approximation of Kirchhoff's voltage law is presented in constraint (2.22). The renewable energy output R_{it}, demand D_{it}, and phase angle β_{it} are usually given as parameters in the transmission constraints.

$$(f_{ijt} - f_{jit}) - B_{ijt}(\beta_{it} - \beta_{jt}) = 0 \qquad \forall\ (i, j) \in A,\ t \in T \qquad (2.22)$$

$$\beta_{it} \text{ unrestricted,} \qquad\qquad \forall\ i \in N,\ t \in T \qquad (2.23)$$

where

β_{it}: a phase angle at interconnected bus i
B_{ijt}: susceptance of an transmission line (i, j)

The system voltage and transformer tap limits are shown in constraint (2.24) and (2.25), respectively.

$$\mathbf{V}^{min} \leq \mathbf{V} \leq \mathbf{V}^{max}, \qquad\qquad (2.24)$$

$$\mathbf{B}^{min} \leq \mathbf{B} \leq \mathbf{B}^{max}, \qquad\qquad (2.25)$$

where

$\mathbf{V}$: system voltage vector
$\mathbf{B}$: transformer tap vector
$\mathbf{V}^{min}, \mathbf{V}^{max}$: system voltage lower and upper limit vector
$\mathbf{B}^{min}, \mathbf{B}^{max}$: transformer tap lower and upper limit vector

2.3.5 Emission Constraints

Environmental factor is one of operation considerations and usually addressed as a system level or regional emission limit in general. The emission control is mainly executed on these emission gases, i.e. CO_2, SO_2, NO_x. Also, the allowable emission amount highly depends on the fuel type of generating unit, for example, a coal-burning electric generating unit has a higher emission level than a gas-turbine unit. A system level emission limit over a planning horizon [FSL05b] is formulated as

$$\sum_{g \in G} \sum_{t \in T} \left(F_g^e(p_{gt})u_{gt} + SU_g^e v_{gt} + SD_g^e w_{gt} \right) \leq E^{max}, \qquad (2.26)$$

where

$F_g^e(\cdot)$: emission function of unit g
SU_g^e: startup emission of unit i at time t
SD_g^e: shutdown emission of unit i at time t
E^{max}: system emission limit

This constraint is applied to one emission gas and the emission function may vary according to the fuel type of generating units. In addition, this constraint can be tailored for regional emission limit based on the location area of generating units.

2.3.6 Unserved Energy Constraint

In some circumstances, load loss is allowed to occur and may come with unserved energy penalty reflected in the objective function. While the unserved energy constraint imposes a performance bounding to control the expected total load losses within an expected loss allowance.

$$E(\sum_{i \in N} \delta_{it}) \leq \varepsilon_t, \ \forall \, t \in T \tag{2.27}$$

where

$E(\cdot)$: the expectation of load loss in a power system
  ε_t: loss allowance for time t

2.3.7 Reactive Power Constraints

Relative to real power generation, this subsection briefly introduce reactive power generation in current system operating, including generation limit, load bus balance and operating reserve requirement [].

$$Q_g^{min} u_{gt} \leq q_{gt} \leq Q_g^{max} u_{gt} \qquad\qquad \forall \, g \in G, \, t \in T \tag{2.28}$$

$$\sum_{g \in G} Q_g^{max} u_{gt} \geq D_t^Q, \qquad\qquad \forall \, t \in T \tag{2.29}$$

load bus balance $\hspace{6cm}$ (2.30)

where

q_{gt}:   reactive power generation of unit g at time t
Q_g^{min}:   lower limit of reactive power generation of unit g
Q_g^{max}:   upper limit of reactive power generation of unit g
D_t^Q: reactive power flow demand at time t

2.4 Case Studies

This section provides two selected cases to illustrate basic unit commitment problems and their solution analyses. Both cases are based on a modified 7-bus system, which are taken from Reference [HZW14]. The test system includes 4 generators, 1 wind farm, and 10 transmission lines with given capacities, shown on Fig. 2.2. The bus parameters corresponding to generating units are listed on Table 2.1. The generating unit parameters and their bid prices are given on Table 2.2. The transmission line parameters are given in Table 2.3. Here, line congestion is not considered in both case studies. The daily forecasted Loads are shown in Fig. 2.3 and the wind energy output is in Fig. 2.4. All models can be coded in C++ and solved by commercial solvers like CPLEX.

Here are two UC cases discussed as follow:

- Case 1: Joint energy and ancillary service optimization
- Case 2: Security-Constrained unit commitment with transmission contingency

Based on the given system, the case studies do not consider the impacts of transformers, phase shifter for MW control as well as contingency, i.e. generator outages, line outage.

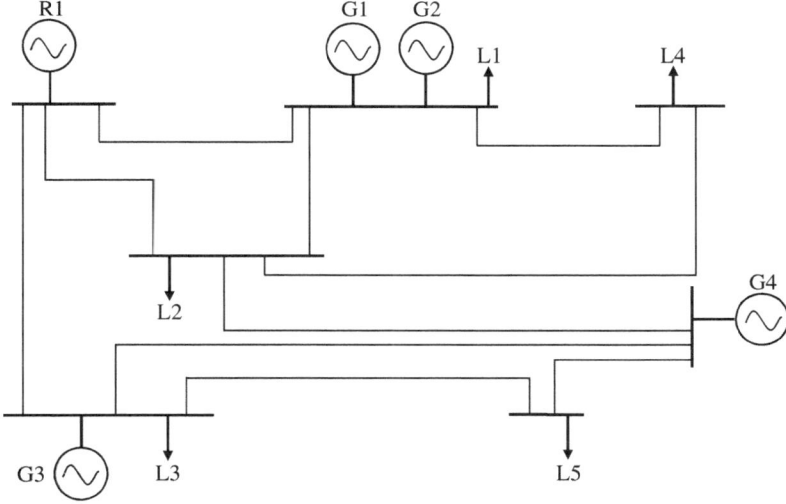

Fig. 2.2 The 7-bus system

Table 2.1 Bus parameters

Bus ID	Type	Unit ID	Gen Capacity (MW)	Spin Reserve limits (MW)	ES Cap. (MW)
B1	Wind	R1	100	–	20
B2	Coal	G1	90	10	20
B2	Coal	G2	60	–	–
B3	–	–	–	–	–
B4	Gas	G3	100	–	10
B5	–	–	–	–	–
B6	Coal	G4	90	–	–
B7	–	–	–	–	–

[a] The symbol, '−', represents no generation unit available at a corresponding bus

Table 2.2 Generator parameters and costs

	G1	G2	G3	G4
Min-ON (h)	2	1	2	2
Min-OFF (h)	2	2	2	1
Ramp-Up Rate(MW/h)	30	15	60	15
Ramp-Down Rate (MW/h)	15	15	60	15
P^{min} (MW)	20	10	20	15
P^{max} (MW)	90	50	90	60
S^{max} (MW)	15	10	15	10
Startup ($)	500	500	800	300
Shutdown ($)	500	500	800	300
Fuel Cost a ($)	6.78	6.78	31.67	10.15
Fuel Cost b ($/MWh)	12.888	12.888	26.244	17.820
Fuel Cost c ($/MWh^2)	0	0	0	0

2.4.1   Case 1: Joint Energy and Ancillary Service Optimization

This case focuses on the co-optimization of energy and ancillary service at a same planning horizon. This energy-reserve co-optimization aims to clear both markets simultaneously in a least-cost way. Although energy and spinning reserve come from the same physical resources, the same amount of electricity provided have different prices between energy market and ancillary service market. The problem is formulated in a two-stage mixed integer linear program. The UC schedule is modeled

Table 2.3 Transmission line parameters

Line ID	From	To	Flow capacity (MW)	Voltage (V)	Susceptance
L1	B1	B2	50	500	1
L2	B1	B3	160	500	1
L3	B1	B4	80	500	1
L4	B2	B3	100	500	1
L5	B2	B5	50	500	1
L6	B3	B5	30	500	1
L7	B3	B6	100	500	1
L8	B4	B6	50	500	1
L9	B4	B7	60	500	1
L10	B6	B7	50	500	1

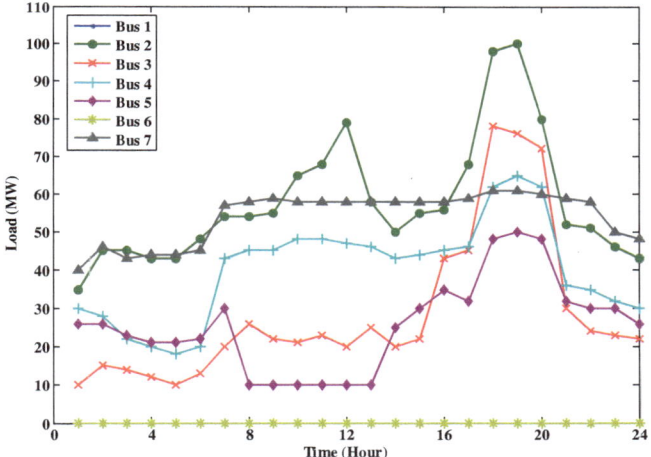

Fig. 2.3 Hourly loads on 7 buses

in the first stage, while both economic dispatch and spinning reserve are scheduled in the second stage.

The length of the planing horizon is 24 h and the forecasted wind energy output is given in one scenario. The wind farm is located at Bus 1 with a generating capacity of 100 MW. The hourly wind energy output was truncated in the range of [5, 80] MW with assumptions of a minimum production output and a maximum production output. Therefore, index sets for Case 1 are shown below and the hourly wind energy outputs are plotted in Fig. 2.4.

$G = 4$ Generators
$T = 24$ Hours

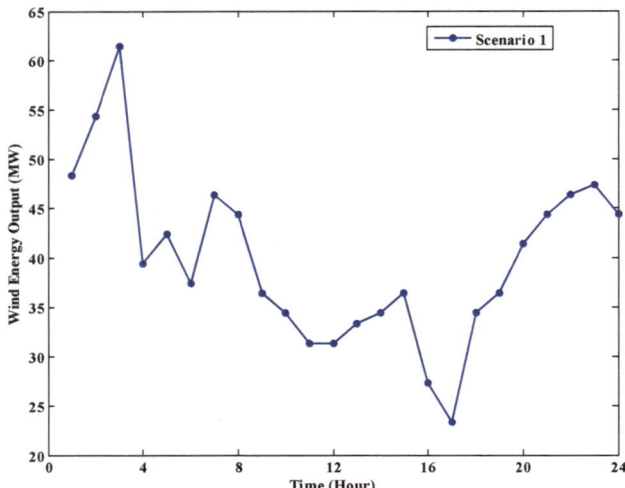

Fig. 2.4 Case 1: hourly wind energy output

$N = 7$ Buses
$S = 1$ Scenario
$|\mathscr{A}| = 10$ Transmission Lines

Then the determinist UC problem for energy and ancillary service is formulated.

$$\min \sum_{g \in G} \sum_{t \in T} (SU_{gt} v_{gt} + SD_{gt} w_{gt}) + \sum_{t \in T} \sum_{g \in G} [(b_{gt} p_{gt} + a_{gt} u_{gt}) + (b'_{gt} s_{gt} + a'_{gt} u_{gt})]$$

$$+ VOLL \sum_{t \in T} \sum_{i \in N} \Delta_{it}$$

s.t. The first-stage constraints

$$u_{gt} - u_{g(t-1)} \leq u_{g\tau}, \quad \forall g \in G,\ t \in T, \tau = t, \ldots, min\{t + L_g - 1\}$$

$$u_{g(t-1)} - u_{gt} \leq 1 - u_{g\tau}, \quad \forall g \in G,\ t \in T, \tau = t, \ldots, min\{t + l_g - 1\}$$

$$v_{gt} \geq u_{gt} - u_{g(t-1)}, \quad \forall g \in G,\ t \in T$$

$$w_{gt} \geq -u_{gt} + u_{g(t-1)}, \quad \forall g \in G,\ t \in T$$

$$u_{gt}, v_{gt}, w_{gt} \in \{0, 1\}, \quad \forall g \in G,\ t \in T$$

The second-stage constraints

$$P_g^{min} u_{gt} \leq p_{gt} \leq P_g^{max} u_{gt}, \quad \forall g \in G,\ t \in T$$

$$- RD_g \leq p_{gt} - p_{gt-1} \leq RU_g, \quad \forall g \in G,\ t \in T$$

$$0 \leq s_{gt} \leq S_g^{max}, \quad \forall g \in G,\ t \in T$$

$$p_{gt} + s_{gt} \leq P_g^{cap} u_{gt}, \quad \forall g \in G,\ t \in T$$

Table 2.4 Objective value and unit commitment for 7-bus system

Objective value	Unit ID	Hour (1–24)
$60615.6	G1	1 1 1 1 1 1 1 1 1 1 1 1 1 1 1 1 1 1 1 1 1 1 1 1
	G2	1 1 1 1 1 1 1 1 1 1 1 1 1 1 1 1 1 1 1 1 1 1 1 1
	G3	0 0 0 0 0 0 0 0 0 0 0 0 0 0 0 0 1 1 1 1 0 0 0 0
	G4	0 0 0 0 0 0 0 0 0 1 1 1 1 1 1 1 1 1 1 1 1 1 1 1

$$\sum_{g \in G_i} s_{gt} \geq RS_{it}, \quad \forall i \in N, t \in T$$

$$\sum_{(i,j) \in A_i^+} f_{ijt} - \sum_{(j,i) \in A_i^-} f_{jit} - \sum_{g \in G_i} (p_{gt} + s_{gt}) - \Delta_{it} = W_{it} - D_{it}, \quad \forall i \in N, t \in T$$

$$(f_{ijt} - f_{jit}) - B_{ijt}(\beta_{it} - \beta_{jt}) = 0, \quad \forall (i, j) \in A, t \in T$$

$$p_{gt}, s_{gt} \geq 0, \quad \forall g \in G, t \in T$$

$$\Delta_{it} \geq 0, \quad \forall i \in N, t \in T$$

$$f_{ijt}, \quad \forall (i, j) \in \mathscr{A}, t \in T$$

We can obtain the computational results using the solver CPLEX, in which the Brand-and-Cut-and-Price algorithm is used to solve mixed integer linear programs. The total generation cost for this recommended UC schedule is $60615.6. Table 2.4 lists the objective value and the optimal UC schedule according to the forecasted (known) hourly wind energy outputs and loads. Figures 2.5 and 2.6 are the optimal generator dispatches and spinning reserve levels, respectively.

Without consideration of line congestion, load-shedding loss can be resulted from the physical generation conditions, such as generation limits or ramping constraints. The solution shows no loss occurs under this wind scenario. Therefore, the current generation capacities and ancillary service requirements are able to provide power balance in this system.

2.4.2 Case 2: SCUC with Transmission Contingency

This case focuses on the N-1 reliable DC optimal dispatch under transmission line outage. This problem bases on the Case 1's model and further considers the impacts of transmission contingency on operation scheduling. The model remains a mixed integer linear program and includes the transmission flow capacity constraint (2.31) subject to a line outage during a period $[t, t + a]$.

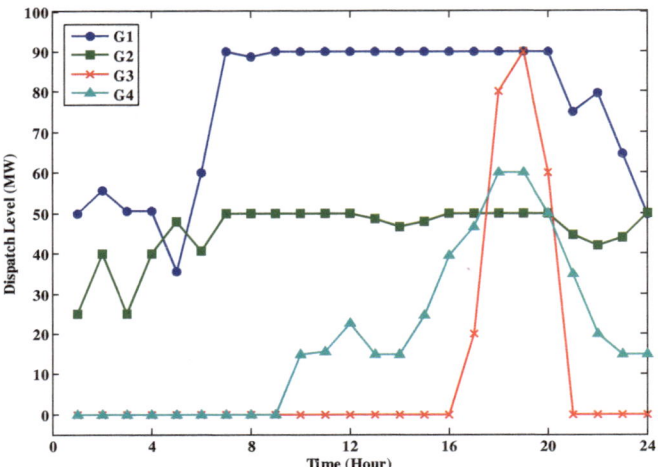

Fig. 2.5 7-bus system: dispatch level for each generator

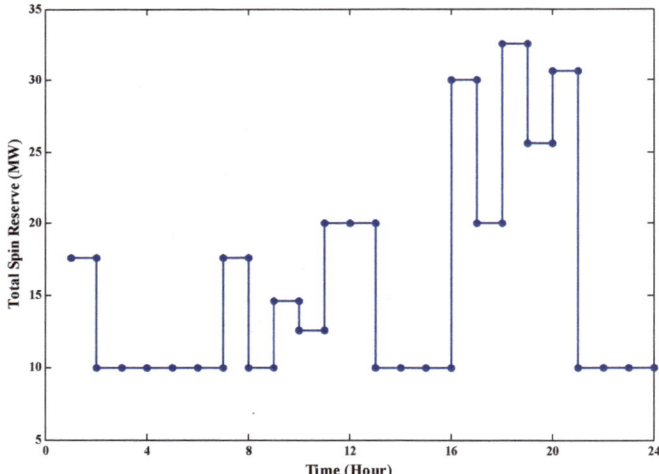

Fig. 2.6 7-bus system: total spin reserve

$$- F_{ij}^{max} \alpha_{ijt} \leq f_{ijt} \leq F_{ij}^{max} \alpha_{ijt}, \qquad \forall\, (i, j) \in A,\ t \in T \qquad (2.31)$$

where

α_{ijt}: Binary parameter, if $\alpha_{ijt} = 1$, line outage occurs between bus i and bus j at time t; otherwise, $\alpha_{ijt} = 0$.

Note that the practical method to deal with transmission line outage is not limited to UC operation scheduling, including common transmission switching, whereas the state of transmission element (line or transformer), voltage and phase angle are fully taken into account in transmission switching. For the simplicity of case, here we do not consider such factors except transmission line.

Case 2 uses the same 7-bus system and shares the same parameters with Case 1. Assuming that the occurrence of line outage can be predicted in advance, only one line outage occurs in line $(4, 7)$ at 11 am. During the line outage, the number of available transmission lines is reduced to 9 and the flow capacity $F^m_{47(11)}ax$ becomes zero. The deterministic UC model for Case 2 is shown as follow.

$$\min \sum_{g \in G} \sum_{t \in T} (SU_{gt} v_{gt} + SD_{gt} w_{gt}) + \sum_{t \in T} \sum_{g \in G} [(b_{gt} p_{gt} + a_{gt} u_{gt}) + (b'_{gt} s_{gt} + a'_{gt} u_{gt})]$$

$$+ VOLL \sum_{t \in T} \sum_{i \in N} \Delta_{it}$$

s.t. The first-stage constraints

$$u_{gt} - u_{g(t-1)} \leq u_{g\tau}, \quad \forall g \in G, \, t \in T, \tau = t, \ldots, min\{t + L_g - 1, |T|\}$$

$$u_{g(t-1)} - u_{gt} \leq 1 - u_{g\tau}, \quad \forall g \in G, \, t \in T, \tau = t, \ldots, min\{t + l_g - 1, |T|\}$$

$$v_{gt} \geq u_{gt} - u_{g(t-1)}, \quad \forall g \in G, \, t \in T$$

$$w_{gt} \geq -u_{gt} + u_{g(t-1)}, \quad \forall g \in G, \, t \in T$$

$$u_{gt}, v_{gt}, w_{gt} \in \{0, 1\}, \quad \forall g \in G, \, t \in T$$

The second-stage constraints

$$P^{min}_g u_{gt} \leq p_{gt} \leq P^{max}_g u_{gt}, \quad \forall g \in G, \, t \in T$$

$$-RD_g \leq p_{gt} - p_{gt-1} \leq RU_g, \quad \forall g \in G, \, t \in T$$

$$0 \leq s_{gt} \leq S^{max}_g, \quad \forall g \in G, \, t \in T$$

$$p_{gt} + s_{gt} \leq P^{cap}_g u_{gt}, \quad \forall g \in G, \, t \in T$$

$$\sum_{g \in G_i} s_{gt} \geq RS_{it}, \quad \forall i \in N, \, t \in T$$

$$\sum_{(i,j) \in A^+_i} f_{ijt} - \sum_{(j,i) \in A^-_i} f_{jit} - \sum_{g \in G_i} (p_{gt} + s_{gt}) - \Delta_{it} = -D_{it}, \quad \forall i \in N, \, t \in T$$

$$-F^{max}_{ij} \alpha_{ijt} \leq f_{ijt} \leq F^{max}_{ij} \alpha_{ijt}, \quad \forall (i, j) \in A, \, t \in T$$

$$p_{gt}, s_{gt} \geq 0, \quad \forall g \in G, \, t \in T$$

$$\Delta_{it} \geq 0, \quad \forall i \in N, \, t \in T$$

$$f_{ijt}, \quad \forall (i, j) \in \mathscr{A}, \, t \in T$$

Because of the disruption of line $(4, 7)$ at 11 am, the transmission flows in the network would be changed as well as the dispatch levels for some specific generating units. Table 2.5 shows that the objective value for UC with a line outage is increased to \$336, 438, in which 76.6% of costs come from the loss penalty. When the line outage happens, the new line capacities are not able to satisfy the surge in flow and line congestions also occur between some buses. Therefore, all units are required

Table 2.5 Objective value and unit commitment for 7-bus system with line outage

Objective value	Unit ID	Hour (1–24)
$336, 438	G1	1 1 1 1 1 1 1 1 1 1 1 1 1 1 1 1 1 1 1 1 1 1 1 1
	G2	1 1 1 1 1 1 1 1 1 1 1 1 1 1 1 1 1 1 1 1 1 1 1 1
	G3	1 1 1 1 1 1 1 1 1 1 1 1 1 1 1 1 1 1 1 1 1 1 1 1
	G4	1 1 1 1 1 1 1 1 1 1 0 1 1 1 1 1 1 1 1 1 1 1 1 1

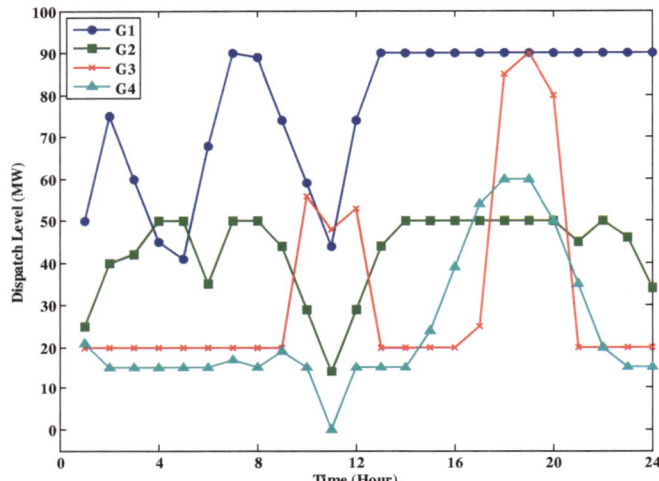

Fig. 2.7 7-bus system: dispatch level for each generator

online and try to meet the local demands first so as to mitigate line congestions. Meanwhile, the line outage forces G4 shut down at 11 am since the outflow of Bus 7 would be terminated.

Figure 2.7 describes the dispatch levels for each generator. Compared to the generation outputs in normal state (Fig. 2.5), these dispatch levels are more fluctuating to accommodate the flow changes. Also, in this case, ramp up/down capabilities regarding online units appear more important to adopt sudden changes in power systems.

Figure 2.8 shows the total spinning reserve in the whole system. Apparently, the overall reserve level is much higher than that of normal state and also the reserve level changes have higher frequency. In the normal state, there is no load-shedding loss within 24 h. However, the line outage results in load-shedding losses gathering at Bus 7 over on-peak hours (Fig. 2.9). Meanwhile, the line outage leading to sudden power supply changes can also trigger losses at other buses, e.g. Bus 3 and Bus 7.

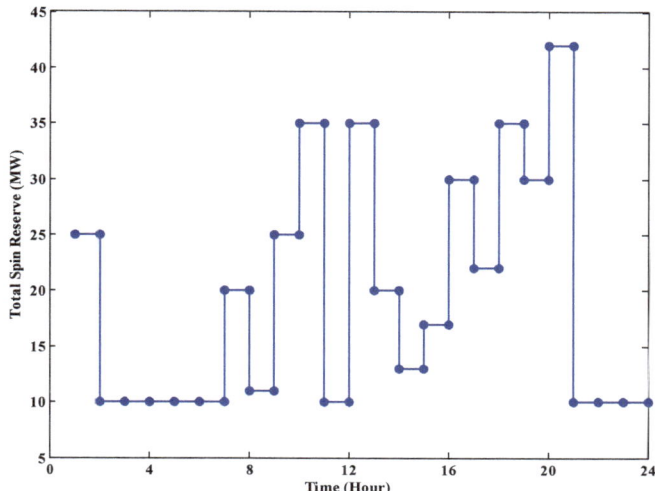

Fig. 2.8   7-bus system: total spin reserve

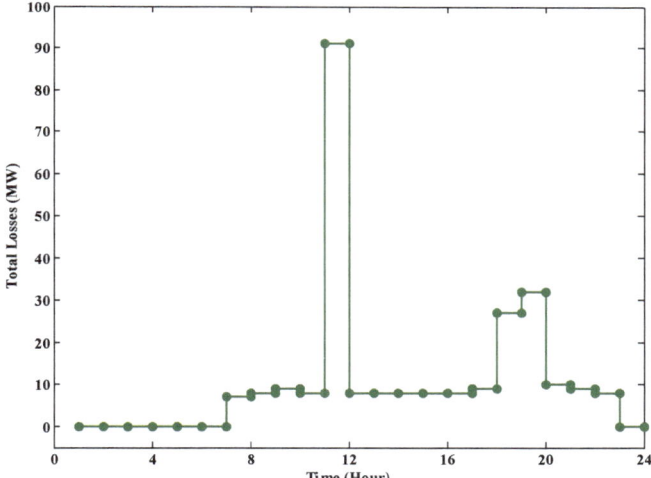

Fig. 2.9   7-bus system: total load-shedding losses

In fact, the unexpected line outage is a very serious contingency event, and thus, it's more applicable to arrange the forced line outage for transmission maintenance to mitigate an unexpected event. What's more, the ISOs/RTOs execute some operating reserves (non-spinning reserve or contingency reserve) to remove the transmission violation, not limited to adjusting tap transformers, phase shifters, predetermined dispatch levels and loads within given time limits.

2.5   Solution Approaches for Deterministic Unit Commitment

We mentioned the unit commitment problems which mainly consider physical generating requirements and power balance as a classic unit commitment problem. In the absence of uncertainty, the classic UC problem is modified to implement some hard operation requirements, i.e. must ON/OFF [OS92, FGL09a, FGL11], operating reserve [SNS01], maintenance [FSL07], emissions [Gje96, FSL05a]. These studies make the classical UC models become more realistic and applicable.

According to the nature of formulated UC problems, there are several common solution approaches for deterministic unit commitment summarized as follow.

- Priority list, including evolutionary programming,
- Dynamic programming,
- Mixed integer linear programming (e.g. Lagrange relaxation method, decomposition method), and
- Heuristics methods.

The above solution approaches have been applied to solve UC problems in the study and the reality. Priority list is one of initial solution methods and dynamic programming is also widely used to obtain UC schedule and optimal generation costs. Mixed integer linear programming has been employed in recent years as the most efficient solution optimization techniques to solve classical UC problems, which will be introduced in Sect. 2.5.3 with more details. Heuristics methods and evolutionary programming have been attempted to solve deterministic UC problems. However, their applications are limited to deterministic cases since they may have lower computational performance when the UC problems face a large-scale power system, and yet both methods can't guarantee for solution optimality. A overview regarding these two solution approaches refers to [Zhu09] for interests.

2.5.1   Priority List

For a generic priority list method, generators are committed in ascending order of the fuel cost so that the most economic base load units are committed first and the most costly units are scheduled last. The priority list method has a very fast computation process, but it is highly heuristic and only generate schedules with relatively high operation cost [SSUF03]. For solving simple UC examples using priority list, the interested reader is referred to [Zhu09, WW12] for more examples.

The usage of priority list is very simple, but is restricted to the basic economic dispatch constraints. This method thus has been extended to accommodate more complicating constraints. The main function of priority list becomes to generate initial solutions due to fast computation speed. Then the initial solutions will go through the improvement process with fast heuristic methods, and eventually one can obtain an economic dispatch schedule and the total generation cost.

In the recent studies, priority list is collaborated with evolutionary algorithms to solve UC problems. Some developed solution approaches provide attempts to solve basic UC problems, such as Evolution Programming, Hybrid Evolution Programming, Gbest based Artificial Bee Colony (ABC) optimization algorithms, Particle Swarm Optimization (PSO) and Differential Evolution technique [GR12a, GR12b].

One of proposed heuristics-based evolutionary algorithm is to evolve an initial population made of good solutions which is obtained by priority list method. Whereas the evolution is characterized by the elimination of the less fit, the survival of the fittest, a reproduction ability based on the fitness, and the genetic operators: crossover, mutation and time-window swap [SC05].

In addition, a hybrid ant system/priority list method is to cooperate the priority list method with the feature of ant system [WCN+09]. The priority list method gives a set of heuristics to be used for UC committing process under the operating constraints. Meanwhile, the ant system can gain the benefit of using a set of heuristic rules provided by the priority list method as directional bias information for improving its evolving process.

What's more, a study proposes an advanced quantum-inspired evolutionary unit commitment algorithm to develop a new searching initialization method based on unit priority list and a special Q-bit expression, which ensures the diversity in the initial search area for improving the efficiency of solution searching. Considering any prior knowledge of UC problem and the characteristics of the generator units, the evolutionary optimization process can be initialized better and carried on by a group-search for QEA-UC [CYW11].

2.5.2 Dynamic Programming

Dynamic programming (DP) is one of main solution techniques to optimize the thermal unit commitment schedule. Dynamic programming with an implicit enumeration approach is a common solution process to solve UC subproblems. Considering an example, there are n generators in a power system, so it has $2^n - 1$ ON-OFF statuses for determining an optimal UC solution. Using DP, it will go through all possible combinations and then pick the best solution(s). The computation times are also increased exponentially, thus the DP applications can't be easily applied in large-scale power systems due to its computational performance.

Some DP introductions with UC applications are clearly given in [Zhu09, WW12]. Generally, dynamic programming is not an unique method employed to produce unit commitment schedules on the whole system. In fact, DP remains its own computational advantages as many studies have proposed DP integrated with other strategies or methods, such as

- Priority list [HHWS88, SPR87, LST+97]
- Lagrangian relaxation method [WSK+95, LS05a, FSL05b]
- Artificial neural network algorithm [OS92]
- Artificial intelligence technique [WS93]
- Expert system [SNS01]
- Branch and bound algorithm [Che08]

The combination of DP with other techniques aims to improve the computation performance. Particularly, within [WSK+95, LS05a, FSL05b, GNLL97], DP is used to solve specific UC subproblems in which the objective is required to determine the optimal unit status cross hours. For the detailed solution process through the DP-Lagrangian relaxation method, one can refer Sect. 2.5.4.3.

2.5.3 Mixed Integer Linear Programming

Compared to other mentioned solution approaches, MILP is the most promising solution technique and has been successfully applied in UC problems. The classic unit commitment problem in abstract form is shown on the following mixed 0–1 linear program (2.32a):

$$[\mathbf{P}] : \min \mathbf{c}_1^T \mathbf{x} + \mathbf{c}_2^T \mathbf{y} \tag{2.32a}$$

$$\text{s.t. } \mathbf{A}_1 \mathbf{x} = \mathbf{b}_1 \tag{2.32b}$$

$$\mathbf{A}_2 \mathbf{x} + \mathbf{E} \mathbf{y} = \mathbf{b}_2 \tag{2.32c}$$

$$\mathbf{x} \in \{0, 1\}^{n_1} \tag{2.32d}$$

$$\mathbf{y} \in \mathbb{R}_+^{n_2} \tag{2.32e}$$

where $\mathbf{c}_1 \in \mathbb{R}^{n_1}, \mathbf{c}_2 \in \mathbb{R}^{n_2}, \mathbf{b}_1 \in \mathbb{R}^{m_1}, \mathbf{b}_2 \in \mathbb{R}^{m_2}, \mathbf{A}_i \in \mathbb{R}^{n_1 \times m_i} (i = 1, 2), \mathbf{E} \in \mathbb{R}^{n_2 \times m_2}$, and m_1, m_2 are scalars.

The mixed integer program contains an integer variable vector, $\mathbf{x}$, and a continuous variable vector, $\mathbf{y}$. The set of constraint (2.32b) represents unit commitment constraints only involving binary variables, while the set of sconstraint (2.32c) mainly covers the generation limits, operating reserve, ramping limits and emission constraints.

The main applications of MILP in UC can be extended with helps from two aspects: problem reformulation and algorithm modification, both of which aim to improving solution process as well as achieving optimal solution easier and faster.

As solving UC representations purely through dynamic programming would cause computational issues, the reformulations to UC problem can be completed using MILP. This aspiration of better formulations promotes seeking alternative representations to get rid of some computational obstacles, such as nonlinear structures. For instance, the original fuel cost function is a mixed integer quadratic function of dispatch/production level, but there exists some situations where directly solving this function may lead to solutions hardly reaching the global optima. To reduce the computational burden, the piecewise linear approximation technique is used to obtain an approximated value for generation variables [CA06, ZWPG13]. In addition, the traditional thermal generator constraints regarding to minimum ON/OFF time are reformulated through pure integer programming (i.e. unit commitment and startup/shutdown action constraint); ramping up/down constraints are simplified from the general ramping constraint describing the relationship between ramping and load level; similar to fuel cost function, the emission constraints can be linearized and show in mixed integer linear programs. Other recent studies on reformulations have reported alternative UC reformulations from mixed integer nonlinear programs and how to make MILP approximations more close to real solutions [FGL09b, Jab12, MELR13].

Regarding the second aspect (algorithm modifications), solution algorithms have been being developed for several years and also bring a lot of vitality to the application of MILP in UC problems. The traditional solution algorithms have been tailored and customized in the way of integrating basic solution algorithms with other solution strategies or decomposing original problems to master problem and multiple subproblems, so that they can be more suitable for applying in new developed UC models. As for deterministic UC problems, solving corresponding mixed integer linear programs in the last decade utilized one or two following solution technique(s) for better computational performance.

- Lagrangian relaxation technique [LB99, MMSN05, MMSN06, FLS+09, LS05b, WSF10],
- Benders' Decomposition [GGZ05b, GGZ05a, FSL05b, LS05a, LS05b],
- Branch-and-Cut method [WSF10],
- Augmented Lagrangian relaxation (LR) method and dynamic programming [FSL05b, LS05a],
- Tabu search [MMSN05, MMSN06],
- Hybrid subgradient and Dantzig–Wolfe decomposition approach [FSL05a]

Here, we mainly introduce two of effective solution techniques, i.e. Lagrangian relaxation (LR) and Benders' Decomposition (BD). Since recent UC problems involving uncertainties and their optimization models become more complicated, Lagrangian relaxation and Benders' Decomposition methods work as fundamental solution theories that provide help for developing other advanced solution algorithms.

Taking the benefits from decomposition methods, a large MILP model can be decomposed into smaller subproblem(s) which can be solved by existing solution algorithms easily, so that computation performance is improved.

Generally, a UC original problems can be solved directly via Brand-and-Cut-and-Price algorithms using solver CPLEX. After breaking down the original problem, if the subproblem is a MILP, the Branch-and-Cut method is suitable for solving this subproblem as well as Branch-and-Price method. If the subproblem is a linear program, many well-known linear algorithms can handle it easily. For solving the optimal commitments in master problems, LR and DP together can be applied to solve short-term UC problems; meanwhile, Tabu Search can be used in the attempt to solve a small size of UC. While solving a long-term UC might be still a challenge for current optimization methods, Fu et al. thus proposed a hybrid subgradient and Dantzig–Wolfe decomposition approach to tackle this issue.

2.5.4   *Lagrangian Relaxation*

Lagrangian Relaxation (LR) is a powerful relaxation technique, which is often used to solve UC problems. As many UC problems are complicated by a number of coupling constraints, their original problems can be modeled as (relatively) easy solving Lagrangian problems. More specifically, the problem reformulation is to replace the complicated constraints with penalty terms in the objective function, in which penalty terms are represented by the violation of constraints and their Lagrangian multipliers. In a Lagrangian problem, a lower bound can be obtained for the optimal value of the minimum non-convex UC problem [FLW13].

As an example, the general solution process for the SCUC model is shown on Fig. 2.10. The solution process starts from solving the master problem (MP), in which the constraints namely include unit commitment, economic dispatch, energy reserve, emission limit and unserved energy limit constraints. In the normal case without any outage, if the MP is found to be feasible, the incumbent solution (UC & ED) is passed to the subproblem for network security evaluation (NSE). If the incumbent solution satisfies the transmission requirements, the ED solution continues to be checked for contingency in NSE. If there is any incumbent solution that fails in NSE for both cases, the MP will be resolved for another solution.

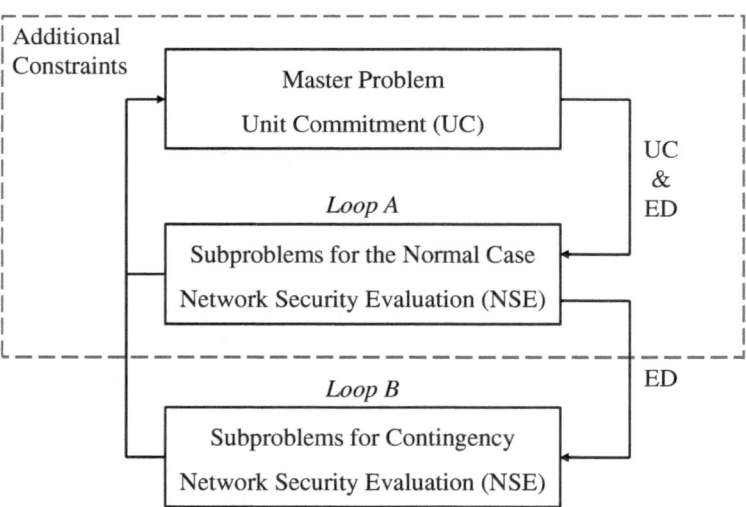

Fig. 2.10 The decomposition approach for SCUC [FLW13]

2.5.4.1 Application of Lagrangian Relaxation in UC Problem

Generally the abstract LR-based UC model can be written in (2.33).

$$[\textbf{LR-OP}] : \min \ \mathbb{F}(x_1, x_2, \ldots, x_{n_1}, y_1, y_2, \ldots, y_{n_2}) \qquad (2.33a)$$
$$\text{s.t.} \ \ \mathbb{H}(x_1, x_2, \ldots, x_{n_1}, y_1, y_2, \ldots, y_{n_2}) \leq \mathbf{d} \quad \breve{} \qquad (2.33b)$$
$$\mathbb{G}_i(x_i, y_i) \leq b_i, \quad \forall \, i \qquad (2.33c)$$
$$\mathbf{x} \in \{0, 1\}^{n_1} \qquad (2.33d)$$
$$\mathbf{y} \in \mathbb{R}_+^{n_2} \qquad (2.33e)$$

where constraints (2.33b) represent a set of coupling constraints, such as reserve requirements, emission constraints, fuel constraints and unserved energy limits, and constraints (2.33c) involve other non-coupling generation constraints, such as minimum ON/OFF constraints, startup/shutdown constraints, generation capacities, ramping limits, spinning/nonspinning constraints and so on.

Here we address the process how to create a Lagrangian problem to solve UC model. We first let non-negative $\breve{}$ denote the Lagrangian multipliers for the system coupling constraints (2.33b). The Lagrangian relaxations of the original problem 2.33 is to move the coupling constraint (2.33b) to the objective function, shown as

$$V D^* = \min \ \mathbb{F}(x_1, x_2, \ldots, x_{n_1}, y_1, y_2, \ldots, y_{n_2})$$
$$+ \ \breve{}(\mathbb{H}(x_1, x_2, \ldots, x_{n_1}, y_1, y_2, \ldots, y_{n_2}) - \mathbf{d}) \qquad (2.34)$$

subject to the unit constraints in (2.33c). When $\breve{}$ is a fixed value, the term $-\breve{}^T \mathbf{d}$ becomes constant and is discarded. Then the LR-based objective function can be decomposed into n_1 subproblems, where each subproblem 2.35 bases on a corresponding generator, shown as follow:

$$[\textbf{LR-SP}] : \min \ \mathbb{F}(x_i, y_i) + \breve{}^T \mathbb{H}_i(x_i, y_i) \qquad (2.35a)$$
$$\text{s.t.} \ \ \mathbb{G}_i(x_i, y_i) \leq b_i, \qquad (2.35b)$$
$$x_i \in \{0, 1\} \qquad (2.35c)$$
$$y_i \in \mathbb{R}_+ \qquad (2.35d)$$

As for solving the decoupled subproblems for each generator, dynamic programming (DP) has been verified as one of effective ways to generate every possible state at each DP stage. Many general discussions of DP can be found in the literature.

In the **LR-SP**, a state space is made up with all possible generator status and then DP will execute searching the best strategy from possible strategies of each stage. Once the generator state x_i and its power dispatch y_i over the planning horizon are determined, we can obtain the objective value for $V D^*$. This is the lower bound of the UC problem and will be used as the dual value. We then examine the relaxed coupling constraints to be satisfied. If these constraints can not be satisfied, the Lagrangian multipliers $\breve{}$ will be updated through another method (e.g., subgradient method).

If they are satisfied, based on the given UC solution, the economic dispatch problem will be solved to determine power dispatch amount on each generator.

$$VP^* = \min \mathbb{F}(\hat{x}_1, \hat{x}_2, \ldots, \hat{x}_{n_1}, y_1, y_2, \ldots, y_{n_2})$$

$$\text{s.t.} \mathbb{H}(\hat{x}_1, \hat{x}_2, \ldots, \hat{x}_{n_1}, y_1, y_2, \ldots, y_{n_2}) \leq \mathbf{d} \tag{2.36}$$

$$\mathbb{G}_i(\hat{x}_i, y_i) \leq b_i, \quad \forall i \tag{2.37}$$

$$y \in \mathbb{R}_+^{n_2} \tag{2.38}$$

The objective value of VP^* is the primal value as the upper bound of the UC problem. We then compare the primal value with the dual value and examine their difference met within the range of duality gap. If the current difference exceeds the duality gap, Lagrange multiplier will be updated until another feasible solution is obtained and the duality gap stays in an acceptable range. So far the LR method has been applied to some specific coupling constraints relaxation, usually for ramping, hydropower generation, transmission network, and emission constraints.

What's more, due to the non-convexity of UC optimization problem, the performance of LR is highly affected by the multipliers and less sufficient to finding a global optimal solution with reasonable convergence speed. Then the augmented Lagrangian method can be applied to deal with the non-convexity in the means of adding quadratic penalty terms to the Lagrangian function. For general UC models, the main difference between Lagrangian Relaxation and Augmented Lagrangian Relaxation exists in the Lagrangian function. In order to improve the convexity of problem, in general we add a quadratic penalty term $-(c/2) \sum_{t \in T} (\sum_{g \in G} p_{gt} u_{gt} - D_t)^2$, which stands for the gap between supply and demand [FSL05b, SYL03].

2.5.4.2   LR Example

To illustrate the implementation of LR in UC problem, we construct a typical UC model to show the LR-based model and its solution process in details. We consider the following UC model with partial prevailing constraints and decompose the original model via the Lagrangian Relaxation method.

$$\min \sum_{g \in G} \sum_{t \in T} (SU_{gt} v_{gt} + SD_{gt} w_{gt}) + \sum_{t \in T} \sum_{g \in G} (b_{gt} p_{gt} + a_{gt} u_{gt}) \qquad .$$

$$\text{s.t.} \, u_{gt} - u_{g(t-1)} \leq u_{g\tau}, \quad \forall g \in G, \, t \in T, \tau = t, \ldots, \min\{t + L_g - 1, \|T\|\} \tag{2.39a}$$

$$u_{g(t-1)} - u_{gt} \leq 1 - u_{g\tau}, \quad \forall g \in G, \, t \in T, \tau = t, \ldots, \min\{t + l_g - 1, \|T\|\} \tag{2.39b}$$

$$v_{gt} \geq u_{gt} - u_{g(t-1)}, \quad \forall g \in G, \, t \in T \tag{2.39c}$$

$$w_{gt} \geq -u_{gt} + u_{g(t-1)}, \quad \forall g \in G, \ t \in T \tag{2.39d}$$

$$u_{gt}, v_{gt}, w_{gt} \in \{0, 1\}, \quad \forall g \in G, \ t \in T \tag{2.39e}$$

$$\sum_{g \in G} p_{gt} u_{gt} = D_t + \Delta_t, \quad \forall t \in T \tag{2.39f}$$

$$\sum_{g \in G} s_{gt} u_{gt} \geq RS_t, \quad \forall t \in T \tag{2.39g}$$

$$P_g^{min} u_{gt} \leq p_{gt} \leq P_g^{max} u_{gt}, \quad \forall g \in G, \ t \in T \tag{2.39h}$$

$$p_{gt} - p_{gt-1} \leq P_g^{min}(2 - u_{gt} - u_{g(t-1)}) + RU_g(1 + u_{g(t-1)} - u_{gt}),$$
$$\forall g \in G, t \in T \tag{2.39i}$$

$$p_{gt-1} - p_{gt} \leq P_g^{min}(2 - u_{gt} - u_{g(t-1)}) + RD_g(1 - u_{g(t-1)} + u_{gt}),$$
$$\forall g \in G, t \in T \tag{2.39j}$$

$$\sum_{g \in G} \sum_{t \in T} \left(F_g^e(p_{gt}) u_{gt} + SU_g^e v_{gt} + SD_g^e w_{gt} \right) \leq E^{max} \tag{2.39k}$$

$$p_{gt}, s_{gt} \geq 0, \quad \forall g \in G, \ t \in T \tag{2.39l}$$

From the given UC model, all constraints are categorized with the same features into separable constraints, i.e. (2.39a)–(2.39e), (2.39h)–(2.39j), and coupling constraints i.e. (2.39f), (2.39g) and (2.39k). Since these coupling constraints have the common feature that all units are aggregated in one constraint for operational requirement. In the consideration of system-level operation, if one generation variable get changed, other generation variables will be affected simultaneously. According to the LR framework, these coupling constraints are relaxed and placed in the objective function associated with Lagrangian multipliers. In doing so, we can construct a Lagrangian function for this UC problem as follows:

$$L(v_{gt}, w_{gt}, u_{gt}, p_{gt}, \lambda_t^b, \lambda_t^r, \lambda^e)$$
$$= \sum_{g \in G} \sum_{t \in T} (SU_{gt} v_{gt} + SD_{gt} w_{gt}) + \sum_{t \in T} \sum_{g \in G} (b_{gt} p_{gt} + a_{gt} u_{gt})$$
$$- \sum_{t \in T} \lambda_t^b \sum_{g \in G} p_{gt} u_{gt} - \sum_{t \in T} \lambda_t^r \sum_{g \in G} s_{gt} u_{gt}$$
$$- \lambda^e \sum_{g \in G} \sum_{t \in T} \left(F_g^e(p_{gt}) u_{gt} + SU_g^e v_{gt} + SD_g^e w_{gt} \right) \tag{2.40}$$

This Lagrangian function of UC problem is subject to separable constraints (2.39a)–(2.39e), (2.39h)–(2.39j), based on each individual generator.

During the LR decomposition process, when the commitment decision u_{gt} and generation decision p_{gt} are determined for all units over the planning horizon, the objective value of (2.40) in $k + 1^{th}$ iteration can be obtained as the lower bound of original UC problem. Next, we use the current solution $(\hat{u}, \hat{p})$ and check for the coupling constraints. When the current solution is not satisfied with that constraints, the

Lagrangian multiplier $\check{}$ will be updated through the subgradient method. Otherwise, we solve the problem 2.41 with fixed $\hat{\mathbf{u}}$

$$\min \sum_{g \in G} \sum_{t \in T} (SU_{gt}\hat{v}_{gt} + SD_{gt}\hat{w}_{gt}) + \sum_{t \in T} \sum_{g \in G} (b_{gt} p_{gt} + a_{gt}\hat{u}_{gt})$$

$$\text{s.t.} \sum_{g \in G} p_{gt}\hat{u}_{gt} = D_t + \Delta_t, \quad \forall t \in T \tag{2.41a}$$

$$\sum_{g \in G} s_{gt}\hat{u}_{gt} \geq RS_t, \quad \forall t \in T \tag{2.41b}$$

$$P_g^{min}\hat{u}_{gt} \leq p_{gt} \leq P_g^{max}\hat{u}_{gt}, \quad \forall g \in G, t \in T \tag{2.41c}$$

$$p_{gt} - p_{gt-1} \leq P_g^{min}(2 - \hat{u}_{gt} - \hat{u}_{g(t-1)}) + RU_g(1 + \hat{u}_{g(t-1)} - \hat{u}_{gt}),$$
$$\forall g \in G, t \in T \tag{2.41d}$$

$$p_{gt-1} - p_{gt} \leq P_g^{min}(2 - \hat{u}_{gt} - \hat{u}_{g(t-1)}) + RD_g(1 - \hat{u}_{g(t-1)} + \hat{u}_{gt}),$$
$$\forall g \in G, t \in T \tag{2.41e}$$

$$\sum_{g \in G} \sum_{t \in T} \left(F_g^e(p_{gt})\hat{u}_{gt} + SU_g^e\hat{v}_{gt} + SD_g^e\hat{w}_{gt}\right) \leq E^{max} \tag{2.41f}$$

$$p_{gt}, s_{gt} \geq 0, \quad \forall g \in G, t \in T \tag{2.41g}$$

and obtain the corresponding solution $\mathbf{p}$ as well as the upper bound of original UC problem. Then check for the difference between the lower bound and upper bound. If the difference is within a specific gap, the UC final solution is obtained. Otherwise, update $\check{}$ again until the optima is found.

2.5.4.3　LR-Based Solution Process

Here we briefly introduce the augmented Lagrangian relaxation integrated dynamic programming approach to solve UC problem within reasonable computation times [FSL05b]. The flow chat for this solution process is shown in Fig. 2.11 and the solution approach is explained as follow.

LR-Based Solution Approach:

- Step 1: Initiate Lagrangian multipliers, to support with power balance equalities, reserve requirements, system fuel limits, system emission limits, and system security constraints (Benders cuts).
- Step 2: Decouple the relaxed problem into several subproblems to represent individual generators (20). Taking the current values of multipliers, apply DP to solve the UC for each unit over a 24-h planning horizon.
- Step 3: Check all power balance, reserve, fuel, and emission constraints as well as Benders cuts produced from the network security check subproblem.

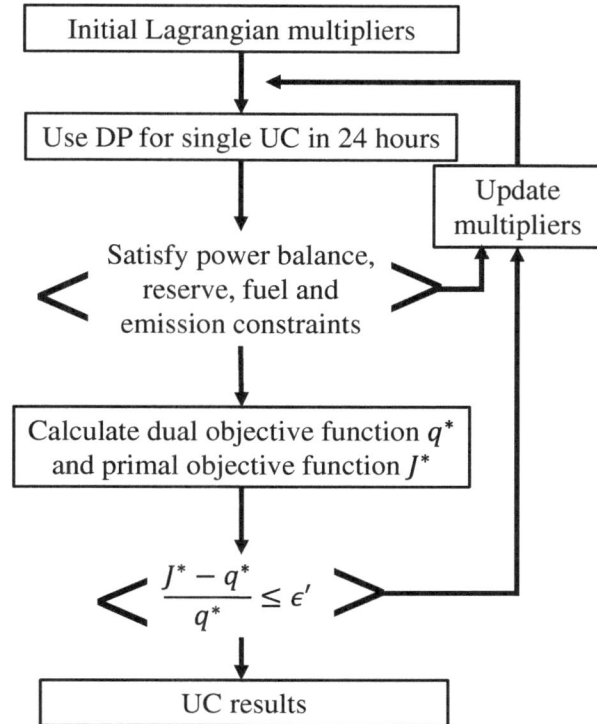

Fig. 2.11 The flow chart of augmented Lagrangian relaxation [FSL05b]

Update multipliers through the subgradient method. Go back to Step 2 if one of constraints cannot be met; otherwise, the solution process will move to Step 4.

• Step 4: Calculate the dual objective in the Lagrangian function and the primal objective (i.e., ED over a 24-h period). Terminate the solving process in the master problem when if the relative duality gap falls in the tolerance; otherwise, keep updating multipliers via the subgradient method, and return to Step 2.

2.5.5 *Benders' Decomposition*

The main use of Benders' decomposition is to decompose an original single large problem into a master problem (MP) and one/multiple smaller subproblems (SP) to alleviate the computational difficulty from directly solving an optimization problem. After decomposition, the algorithm process goes through serval steps: solving MP to get a lower bound, passing its current solutions to SP, solving SP to get a upper bound and then generating Bender's cuts for MP until LB and UB are converged.

As for decomposition, we target to build the subproblem as a linear program (LP) or a convex nonlinear program [CGB06] in that it applies the theory of duality to get a feasible solution, and allow the master problem include all discrete variables, such as binary variables or integer variables. In some cases, one can also keep some of the continuous variables in the master problem according to the needs of master problem and the program structure of subproblem.

2.5.5.1 Principles of Benders' Decomposition

In this section, we consider a MILP-based UC problem and use it as an example to illustrate the procedure of Benders' decomposition. The original UC has two types of decision variables, $\mathbf{x}$ and $\mathbf{y}$, which are vectors of integer and continuous variables. For fixing values of $\mathbf{x}$ variables, the original problem is given by

$$\min\ \{\mathbf{f}(\hat{\mathbf{x}}) + \mathbf{c}_2^T\mathbf{y} \mid \mathbf{E}\mathbf{y} \geq \mathbf{b}_2 - \mathbf{A}_2\hat{\mathbf{x}},\ \mathbf{y} \in \mathbb{R}_+,\ \mathbf{y} \geq 0\}. \tag{2.42}$$

Since the value of function $\mathbf{x}$ is fixed in the objective function and moved out from the function $\mathbf{y}$, the problem (2.42) can be written as follow:

$$\mathbf{f}(\hat{\mathbf{x}}) + \min\ \{\mathbf{c}_2^T\mathbf{y} \mid \mathbf{E}\mathbf{y} \geq \mathbf{b}_2 - \mathbf{A}_2\hat{\mathbf{x}},\ \mathbf{y} \in \mathbb{R}_+,\ \mathbf{y} \geq 0\}, \tag{2.43}$$

where the inner minimization problem is defined to be subproblem (SP).

Let $^-$ denote dual variables (extreme points in a feasible region) associated with the specific constraint, $\mathbf{E}\mathbf{y} \geq \mathbf{b}_2 - \mathbf{A}_2\hat{\mathbf{x}}$. If $\mathbf{y} \in \mathbb{Y}$ is a nonempty polytope, there exists an extreme point for optimal solution in SP. We can further formulate the dual SP as

$$\min\ \{z \mid z \geq (\mathbf{b}_2 - \mathbf{A}_2\hat{\mathbf{x}})^{T-},\ \mathbf{E}^{T-} \leq \mathbf{c}_2,\ ^- \geq 0\}. \tag{2.44}$$

Solving the inner minimization problem means enumerating all extreme points of Y in the subproblem. If there are partial k ($k < Q$) extreme points selected, the MP becomes a relaxed master problem (RMP) with less constraints given by

$$\min\ \{\mathbf{f}(\mathbf{x}) + z \mid \mathbf{x} \in \mathbb{X},\ z \geq (\mathbf{b}_2 - \mathbf{A}_2\mathbf{x})^{T-}{}_j,\ \text{for } j = 1, 2, \ldots, k\}. \tag{2.45}$$

Define $(\bar{x}, \bar{z})$ as an optimal solution to RMP. In this situation with given partial extreme points, $(\bar{x}, \bar{z})$ can only be considered as a feasible solution to the master problem ($k = Q$). To check this optimality condition, we equivalently check if this solution can make the inequality (2.46) at all extreme points hold true.

$$\bar{z} \geq (\mathbf{b}_2 - \mathbf{A}_2\bar{\mathbf{x}})^{T-}{}_j,\ \text{for } j = 1, 2, \ldots, Q \tag{2.46}$$

If the current solution of RMP, $(\bar{x}, \bar{z})$, violates one or partial constraints in SP, an *optimality cut* (2.47) will be imposed to RMP.

$$z \geq (\mathbf{b} - \mathbf{D}\mathbf{y})^T\hat{\mathbf{u}}_{k+1}. \tag{2.47}$$

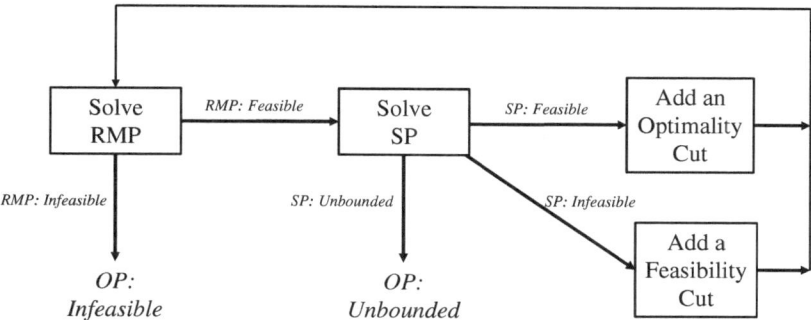

Fig. 2.12 Solution types for master problem and subproblems in Benders' Decomposition

If SP has infeasible solutions, a *feasibility cut* (2.48) will be added to RMP.

$$0 \geq (\mathbf{b} - \mathbf{Dy})^T \hat{\mathbf{u}}_{k+1}. \tag{2.48}$$

During the solving process, MP and SP may experience one or more solution types, shown in the Fig. 2.12. After solving RMP, it may have a feasible solution which will be passed to SP for the next-step solution, or may have an infeasible solution that indicates the original problem to be infeasible. Then the suproblem is solved with three possible cases: feasible, infeasible and unbounded. Based on the solution type of SP, an optimality cut or a feasibility cut will be generated and then added to RMP for next iterations. If the SP has the unbounded case, it also shows that the original problem is unbounded.

To solve a classical MILP problem with L-shaped structure, we outline a traditional Benders' Decomposition algorithm as follow:

▶ **Initialization**: Let $\hat{\mathbf{x}} :=$ initial feasible solution, only solve for the function of x to get the initial LB and then fix $\mathbf{x}$ to solve for UB.
▶ **Step 1**: Solve the RMP, $\min_x \{f(\mathbf{x}) + z | \mathbf{x} \in X, \text{cuts}, z \text{ unrestricted}\}$.
   If RMP is feasible, get solutions $(\bar{}, \bar{z})$ and $LB := f(\bar{\mathbf{x}}) + \bar{z}$; otherwise, the algorithm is terminated.
▶ **Step 2**: Solve the SP, $\max_\mu \{\mathbf{f}(\hat{\mathbf{x}}) + (\mathbf{b_2} - \mathbf{A_2}\hat{\mathbf{x}})^{T^-} | \mathbf{A}^{T^-} \leq \mathbf{c}, ^- \geq 0\}$.
   If SP is feasible, get dual solutions $\hat{}$ and $UB := \mathbf{f}(\hat{\mathbf{x}}) + (\mathbf{b_2} - \mathbf{A_2}\hat{\mathbf{x}})^{T\hat{}}$.
   Add optimality cut $z \geq (\mathbf{b_2} - \mathbf{A_2}\mathbf{x})^{T\hat{}}$ to RMP.
   If SP is infeasible, add feasibility cut $0 \geq (\mathbf{b_2} - \mathbf{A_2}\mathbf{x})^{T\hat{}}$ to RMP.
▶ If $(UB - LB)/UB \leq \epsilon$, the current solution is optimal and the algorithm is terminated.
   If $(UB - LB)/UB > \epsilon$, perform next iteration and go to Step 1.

2.5.5.2 Application of Benders' Decomposition in UC Problem

Based on the above decomposition approach, we can obtain the decomposed UC problems: an integer master problem (BD-MP) and a linear subproblem (BD-SP), which are given by

$$[\textbf{BD-MP}] : LB = \min_{x,\pi} \mathbf{c}_1^T \mathbf{x} + \text{ß} \tag{2.49a}$$

$$\text{s.t.} \quad \mathbf{A}_1 \mathbf{x} = \mathbf{b}_1 \tag{2.49b}$$

$$\mathbf{x} \in \{0, 1\}^{n_1} \tag{2.49c}$$

$$\pi \geq \mathscr{O}(\mathbf{x}) \tag{2.49d}$$

$$0 \geq \mathscr{F}(\mathbf{x}) \tag{2.49e}$$

$$[\textbf{BD-SP}] : UB = \min_y \mathbf{c}_2^T \mathbf{y} \tag{2.50a}$$

$$\text{s.t.} \quad \mathbf{E}\mathbf{y} = \mathbf{b}_2 - \mathbf{A}_2 \hat{\mathbf{x}} \tag{2.50b}$$

$$\mathbf{y} \in \mathbb{R}_+^{n_2} \tag{2.50c}$$

where π is a free variable; constraints (2.49d) and (2.49e) represents a set of optimality cuts and feasibility cuts, respectively.

In the review of decomposition strategies of UC problems the decomposition strategy depending on the types of decision variables has been used a lot, as shown in 2.49 and 2.50.

- Solve the MP with unit commitment and generated cuts;
- Given the current solutions from MP, solve the SP including economic dispatch, operating reserve, emission, transmission, reactive power and unserved energy constraints. Generate Benders' cut(s) according to solution type of SP in current iteration.

Another common strategy of Benders' Decomposition is to solve general security-constrained unit commitment (SCUC) in two operation stages:

- Solve the MP with unit commitment, economic dispatch, operating reserve and emission constraints;
- Given the current solutions from MP, solve the SP only regarding to transmission, reactive power and unserved energy constraints. Check if any network violations occur and generate Benders' cuts.

For both decomposition schemes, the MP includes new generated cuts, and the SP are solved iteratively and checked for convergence. When using the second decomposition scheme, the MP becomes a mixed integer program while the SP is built as a simple linear program and used for meeting network constraints.

From the literature, the network security check is usually arranged in the SP. In particular, the DC network security check focuses on the power flow balance and flow

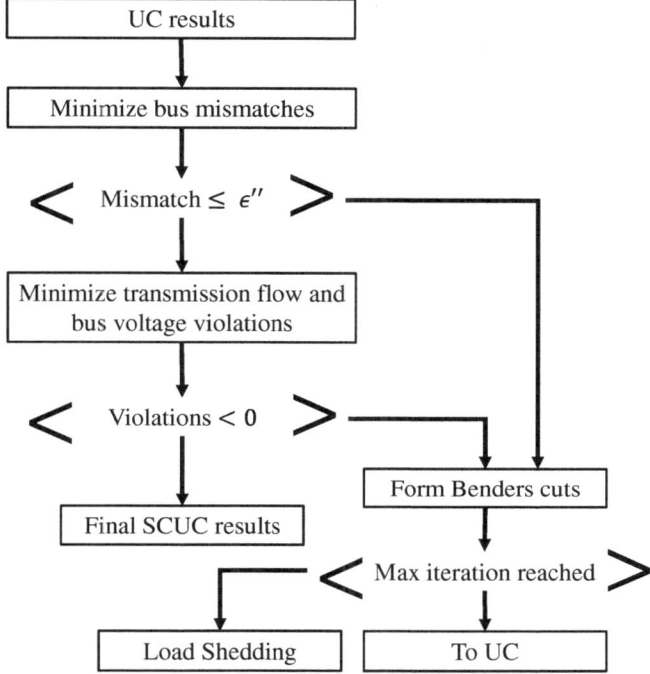

Fig. 2.13 BD-SP: the flow chart of AC network security check [FSL05b]

restrictions on transmission lines. If the DC network constraint is replaced by more complicated AC network constraint, the scheme remains suitable for AC network security check. Because the DC network constraints only consider the power flow balance at a bus and have several limitations, such as ignoring bus voltage violations, feasible distribution of reactive power and interactions between real and reactive power conditions. When the AC network considers such requirements left behind, it is more appropriate to handle them in SP through the security check. The flow chart for a comprehensive network security check in subproblem is shown on Fig. 2.13. This decomposition strategy has also been testified to solve a deterministic large-scale UC problem effectively, i.e. 118 bus system [FSL05b].

2.5.5.3 BD Example

We take the same UC problem shown in Sect. 2.5.4.2 and decompose it using the first strategy of Benders' decomposition. The UC problem is decomposed into a MP and a SP, shown in 2.51 and 2.52. For the second strategy of BD, interested readers can find some explicit examples in [SYL03].

[BD-MP] :

$$\min \quad \sum_{g \in G} \sum_{t \in T} (SU_{gt} v_{gt} + SD_{gt} w_{gt} + a_g u_{gt}) + \pi \tag{2.51a}$$

s.t. $\quad u_{gt} - u_{g(t-1)} \leq u_{g\tau}, \quad \forall g \in G, \, t \in T, \tau = t, \ldots, min\{t + L_g - 1, |T|\}$

$$\tag{2.51b}$$

$$u_{g(t-1)} - u_{gt} \leq 1 - u_{g\tau}, \quad \forall g \in G, \, t \in T, \tau = t, \ldots, min\{t + l_g - 1, |T|\}$$

$$\tag{2.51c}$$

$$v_{gt} \geq u_{gt} - u_{g(t-1)}, \quad \forall g \in G, \, t \in T \tag{2.51d}$$

$$w_{gt} \geq -u_{gt} + u_{g(t-1)}, \quad \forall g \in G, \, t \in T \tag{2.51e}$$

$$u_{gt}, v_{gt}, w_{gt} \in \{0, 1\}, \quad \forall g \in G, \, t \in T \tag{2.51f}$$

$$\pi \geq \mathscr{O}(\mathbf{u}) \tag{2.51g}$$

$$0 \geq \mathscr{F}(\mathbf{u}) \tag{2.51h}$$

[BD-SP] :

$$\min \quad \sum_{t \in T} \sum_{g \in G} b_{gt} p_{gt} \tag{2.52a}$$

s.t. $\quad \sum_{g \in G} p_{gt} = D_t + \Delta_t, \quad \forall t \in T \quad \rightarrow \alpha_t \tag{2.52b}$

$$\sum_{g \in G} s_{gt} \geq RS_t, \quad \forall t \in T \quad \rightarrow \beta_t \tag{2.52c}$$

$$p_{gt} \geq P_g^{min} \hat{u}_{gt}, \quad \forall g \in G, \, t \in T \quad \rightarrow \gamma_{gt} \tag{2.52d}$$

$$p_{gt} \leq P_g^{max} \hat{u}_{gt}, \quad \forall g \in G, \, t \in T \quad \rightarrow \varepsilon_{gt} \tag{2.52e}$$

$$p_{gt} - p_{gt-1} \leq P_g^{min}(2 - \hat{u}_{gt} - \hat{u}_{g(t-1)}) + RU_g(1 + \hat{u}_{g(t-1)} - \hat{u}_{gt}),$$
$$\forall g \in G, t \in T \quad \rightarrow \vartheta_{gt} \tag{2.52f}$$

$$p_{gt-1} - p_{gt} \leq P_g^{min}(2 - \hat{u}_{gt} - \hat{u}_{g(t-1)}) + RD_g(1 - \hat{u}_{g(t-1)} + \hat{u}_{gt}),$$
$$\forall g \in G, t \in T \quad \rightarrow \kappa_{gt} \tag{2.52g}$$

$$\sum_{g \in G} \sum_{t \in T} F_g^e(p_{gt}) \leq E^{max} - \sum_{g \in G} \sum_{t \in T} (SU_g^e \hat{v}_{gt} + SD_g^e \hat{w}_{gt}) \quad \rightarrow \nu$$

$$\tag{2.52h}$$

$$p_{gt}, s_{gt} \geq 0, \quad \forall g \in G, \, t \in T \tag{2.52i}$$

where power balance constraint (2.39f) and spinning reserve requirement (2.39g) are replaced with (2.52b) and (2.52c). The major difference between these two expressions is that the bilinear terms ($p_{gt} u_{gt}, s_{gt} u_{gt}$) are eliminated and only linear terms (p_{gt}, s_{gt}) remain for simplifying the solving process in SP. In addition, the power balance has the same restriction as (2.39f), while the spinning reserve sources are expanded not only from online units but also offline units. In doing so, these modifications can simplify the computation process in SP.

We define several dual variables, such as α_t, β_t, γ_{gt}, ε_{gt}, ϑ_{gt}, κ_{gt}, ν corresponding to constraints (2.52b)–(2.52h), respectively. Then the optimality cut, $\pi \geq \mathcal{O}(\mathbf{u})$, is formed in (2.53) through the dual solution of **BD-SP**.

$$
\pi \geq \sum_{t \in T} \hat{\alpha}_t (D_t + \Delta_t) + \sum_{t \in T} \hat{\beta}_t R S_t + \sum_{g \in G} \sum_{t \in T} \hat{\gamma}_{gt} P_g^{min} u_{gt} + \sum_{g \in G} \sum_{t \in T} \hat{\varepsilon}_{gt} P_g^{max} u_{gt}
$$
$$
+ \sum_{g \in G} \sum_{t \in T} \hat{\vartheta}_{gt} \left[P_g^{min} (2 - \hat{u}_{gt} - \hat{u}_{g(t-1)}) + RU_g (1 + \hat{u}_{g(t-1)} - \hat{u}_{gt}) \right]
$$
$$
+ \sum_{g \in G} \sum_{t \in T} \hat{\kappa}_{gt} \left[P_g^{min} (2 - \hat{u}_{gt} - \hat{u}_{g(t-1)}) + RD_g (1 - \hat{u}_{g(t-1)} + \hat{u}_{gt}) \right]
$$
$$
+ \hat{\nu} [E^{max} - \sum_{g \in G} \sum_{t \in T} (SU_g^e v_{gt} + SD_g^e w_{gt})] \tag{2.53}
$$

This cut is associated with binary variables (u_{gt}, v_{gt}, w_{gt}) and given with incumbent dual values in k^{th} iteration.

2.6 Summary

This chapter introduces basic UC formulations in terms of optimization methods, including objective function and their essential constraints: unit commitment constraints, electricity dispatch, operating reserve constraints, transmission constraints, emission constraints, unserved energy constraints, and reactive power constraints. To address UC problems by optimization approaches, we chose two typical case studies to illustrate how to model UC problems and analyze optimal solutions for better decision making. For the improvement of solution process of UC models, we also provided a overview of solution approaches and summarized their recent development. Particularly, we provided a detailed introduction on the most widely used methods for solving moderate power systems, involving MILP, LR decomposition method and BD decomposition method.

References

[BGC05]   Bouffard F, Galiana FD, Conejo AJ (2005) Market-clearing with stochastic security-part I: formulation. IEEE Trans Power Syst 20(4):1818–1826

[CA06]   Carrion M, Arroyo JM (2006) A computationally efficient mixed-integer linear formulation for the thermal unit commitment problem. IEEE Trans Power Syst 21(3):1371–1378

[CGB06]   Castillo E, Garca-Bertrand RMR (2006) Decomposition techniques in mathematical programming. Springer, Heidelberg

[Che08]   Chen C-L (2008) Optimal wind-thermal generating unit commitment. IEEE Trans Energy Convers 23(1):273–280

[CYW11]   Chung CY, Yu H, Wong K-P (2011) An advanced quantum-inspired evolutionary algorithm for unit commitment. IEEE Trans Power Syst 26(2):847–854

[FGL09a]   Frangioni A, Gentile C, Lacalandra F (2009a) Tighter approximated milp formulations for unit commitment problems. IEEE Trans Power Syst 24(1):105–113

[FGL09b]   Frangioni A, Gentile C, Lacalandra F (2009b) Tighter approximated milp formulations for unit commitment problems. IEEE Trans Power Syst 24(1):105–113

[FGL11]   Frangioni A, Gentile C, Lacalandra F (2011) Sequential lagrangian-milp approaches for unit commitment problems. Int J Electr Power Energy Syst 33(3):585–593

[FLS+09]   Fu Y, Li Z, Shahidehpour M, Zheng T, Litvinov E (2009) Coordination of midterm outage scheduling with short-term security-constrained unit commitment. IEEE Trans Power Syst 24(4):1818–1830

[FLW13]   Fu Y, Li Z, Wu L (2013) Modeling and solution of the large-scale security-constrained unit commitment. IEEE Trans Power Syst 28(4):3524–3533

[FSL05a]   Fu Y, Shahidehpour M, Li Z (2005) Long-term security-constrained unit commitment: hybrid dantzig-wolfe decomposition and subgradient approach. IEEE Trans Power Syst 20(4):2093–2106

[FSL05b]   Fu Y, Shahidehpour M, Li Z (2005) Security-constrained unit commitment with AC constraints. IEEE Trans Power Syst 20(2):1001–1013

[FSL07]   Fu Y, Shahidehpour M, Li Z (2007) Security-constrained optimal coordination of generation and transmission maintenance outage scheduling. IEEE Trans Power Syst 22(3):1302–1313

[GGZ05a]   Guan X, Guo S, Zhai Q (2005) The conditions for obtaining feasible solutions to security-constrained unit commitment problems. IEEE Trans Power Syst 20(4):1746–1756

[GGZ05b]   Guo S, Guan X, Zhai Q (2005) The necessary and sufficient conditions for determining feasible solutions to unit commitment problems with ramping constraints. IEEE Power Eng Soc Gen Meet 1:344–349

[Gje96]   Gjengedal T (1996) Emission constrained unit commitment. IEEE Trans Energy Convers 11(1):132–138

[GNLL97]   Guan X, Ni E, Li R, Luh PB (1997) An optimization-based algorithm for scheduling hydrothermal power systems with cascaded reservoirs and discrete hydro constraints. IEEE Trans Power Syst 12(4):1775–1780

[GR12a]   Govardhan M, Roy R (2012) An application of differential evolution technique on unit commitment problem using priority list approach. In: IEEE international conference on power and energy, pp 858–863

[GR12b]   Govardhan M, Roy R (2012) Evolutionary computation based unit commitment using hybrid priority list approach. In: IEEE international conference on power and energy, pp 245–250

[HHWS88]   Hobbs WJ, Hermon G, Warner S, Shelbe GB (1988) An enhanced dynamic programming approach for unit commitment. IEEE Trans Power Syst 3(3):1201–1205

[HROC01]   Hobbs BF, Rothkopf MH, O'Neil RP, Chao H (2001) The next generation of electric power unit commitment models. Kluwer Academic Publishers, Norwell

[HZW14]   Huang Y, Zheng QP, Wang J (2014) Two-stage stochastic unit commitment model including non-generation resources with conditional value-at-risk constraints. Electric Power Syst Res 116:427–438

[Jab12]   Jabr RA (2012) Tight polyhedral approximation for mixed-integer linear programming unit commitment formulations. IET Gener Transm Distrib 6(11):1104–1111

[LB99]   Lai S-Y, Baldick R (1999) Unit commitment with ramp multipliers. IEEE Trans Power Syst 14(1):58–64

[LS05a]   Li Z, Shahidehpour M (2005) Security-constrained unit commitment for simultaneous clearing of energy and ancillary services markets. IEEE Trans Power Syst 20(2):1079–1088

[LS05b]   Lu B, Shahidehpour M (2005) Unit commitment with flexible generating units. IEEE
            Trans Power Syst 20(2):1022–1034
[LST+97]   Li C, Svoboda AJ, Tseng C-L, Johnson RB, Hsu E (1997) Hydro unit commitment in
            hydro-thermal optimization. IEEE Trans Power Syst 12(2):764–769
[MELR13]   Morales-Espana G, Latorre JM, Ramos A (2013) Tight and compact milp formulation
            for the thermal unit commitment problem. IEEE Trans Power Syst 28(4):4897–4908
[MMSN05]   Mitani T, Mishima Y, Satoh T, Nara K (2005) Security constrains unit commitment by
            lagrangian decomposition and tabu search. In: Proceedings of the 13th international
            conference on intelligent systems application to power systems, pp 440–445
[MMSN06]   Mitani T, Mishima Y, Satoh T, Nara K (2006) Optimal generation scheduling under
            competitive environment. IEEE Int Conf Syst Man Cybern 3:1843–1848
[NLR04]   Ni E, Luh PB, Rourke S (2004) Optimal integrated generation bidding and scheduling
            with risk management under a deregulated power market. IEEE Trans Power Syst
            19(1):600–609
[OS92]   Ouyang Z, Shahidehpour SM (1992) A hybrid artificial neural network-dynamic pro-
            gramming approach to unit commitment. IEEE Trans Power Syst 7(1):236–242
[SC05]   Srinivasan D, Chazelas J (2005) Heuristics-based evolutionary algorithm for solving
            unit commitment and dispatch. In: The 2005 IEEE congress on evolutionary compu-
            tation, 2005, vol 2, pp 1547–1554
[SNS01]   Siu TK, Nash GA, Shawwash ZK (2001) A practical hydro, dynamic unit commitment
            and loading model. IEEE Power Eng Rev 21(5):64–64
[SPR87]   Snyder WL, Powell HD, Rayburn JC (1987) Dynamic programming approach to unit
            commitment. IEEE Trans Power Syst 2(2):339–348
[SSUF03]   Senjyu T, Shimabukuro K, Uezato K, Funabashi T (2003) A fast technique for unit
            commitment problem by extended priority list. IEEE Trans Power Syst 18(2):882–888
[SYL03]   Shahidehpour M, Yamin H, Li Z (2003) Market operations in electric power systems,
            forecasting, scheduling, and risk management. Wiley, New York
[WCN+09]   Withironprasert K, Chusanapiputt S, Nualhong D, Jantarang S, Phoomvuthisarn S
            (2009) Hybrid ant system/priority list method for unit commitment problem with
            operating constraints. In: IEEE international conference on industrial technology, pp
            1–6
[WS93]   Wang C, Shahidehpour SM (1993) Effects of ramp-rate limits on unit commitment
            and economic dispatch. IEEE Trans Power Syst 8(3):1341–1350
[WSF10]   Wu L, Shahidehpour M, Fu Y (2010) Security-constrained generation and transmission
            outage scheduling with uncertainties. IEEE Trans Power Syst 25(3):1674–1685
[WSK+95]   Wang SJ, Shahidehpour SM, Kirschen DS, Mokhtari S, Irisarri GD (1995) Short-
            term generation scheduling with transmission and environmental constraints using an
            augmented lagrangian relaxation. IEEE Trans Power Syst 10(3):1294–1301
[WW12]   Wood AJ, Wollenberg BF (2012) Power generation, operation, and control. Wiley,
            New York
[WWG13b]   Wang Q, Wang J, Guan Y (2013) Price-based unit commitment with wind power
            utilization constraints. IEEE Trans Power Syst 28(3):2718–2726 August
[WWG13c]   Wang Q, Watson J-P, Guan Y (2013) Two-stage robust optimization for N-K
            contingency-constrained unit commitment. IEEE Trans Power Syst 28(3):2366–2375
            August
[YWGZ12]   Yang Y, Wang J, Guan X, Zhai Q (2012) Subhourly unit commitment with feasible
            energy delivery constraints. Appl Energy 96:245–252
[Zhu09]   Zhu J (2009) Optimization of power system operation, vol 49. Wiley, New York
[ZWPG13]   Zheng QP, Wang J, Pardalos PM, Guan Y (2013) A decomposition approach to the two-
            stage stochastic unit commitment problem. Ann Oper Res 210:387–410 November

Chapter 3
Two-Stage Stochastic Programming Models and Algorithms

This chapter discusses the technical and management solution approaches for solving UC problems under uncertainty. There are many recent programs and studies targeted to uncertainty resistance, such as demand response program, energy storage, real-time rescheduling, contingency management, risk measure and control. In recent years great interests has been directed towards reducing the impacts of uncertainty on electrical power system, and the focuses of solving deterministic UC problems are transferred to solving UC problems under uncertainty. One of successful approaches is to apply two-stage stochastic programming to build UC models incorporating system's uncertainties. Also, several commonly used algorithms are introduced because they achieve better computational performance to deal with the large-scale real world problems. Their features and uses in practice are included for reader's comparisons.

3.1 Introduction

The system's uncertainties can be categorized from three aspects: supply-side uncertainty, demand-side uncertainty and transmission uncertainty. The supply-side uncertainty is namely composed of fuel price, unit outage, renewable energy output (e.g. wind- and solar-based generation). The demand-side uncertainty is primarily from local/regional demand changes presenting as daily surges or unexpected surges in demands, due to electricity price, weather condition and social events. The transmission uncertainty is often induced by line outage in contingency, which is hard to be detected before it occurs.

The previous chapter discussed the deterministic UC formulations which only focus on the technical solutions. Regardless of uncertainties on power systems, the deterministic UC optimization models are easy to solve through commercial solvers. However, the uncertainties from internal systems or external systems may occur at any moment and result in disturbances on normal operations and power balance. The factors that can lead to unreliable systems are usually referred to as 'uncertainties'. To

© The Author(s) 2017   49
Y. Huang et al., *Electrical Power Unit Commitment*,
SpringerBriefs in Energy, DOI 10.1007/978-1-4939-6768-1_3

avoid or mitigate the unexpected outcomes, many researchers and decision makers
have proposed many ideas, mathematical models and solution approaches to deal
with different uncertainty issues. The most common uncertainties occurring in power
systems can be summarized as follow.

- Load uncertainty
- Renewable energy outputs (e.g. wind, solar, hydro and geothermal generation)
- Contingency (e.g. generator outages, transmission line outage)

To reduce the impacts of uncertainties on electric power systems, the traditional
method uses spinning reserve, non-spinning reserve and operating reserve to handle the majority of common uncertainties with substantial costs. Although all these
reserves are able to resolve most operational challenges caused by uncertainties,
reserve levels are in fact overestimated usually in normal status while underesti-
mated in extreme cases. A lot of generation resources were 'waste' to safeguard
electric power systems. Therefore, many studies explored and developed the traditional UC models with considerations of uncertainties on two branches: Stochastic
Optimization and Robust Optimization.

Because the development of UC using mathematical programming techniques
is fast growing, particularly mixed integer linear programming (MILP) with board
applications. This section thus focuses on an introduction of two-stage stochastic programming approach to deal with supply-side and demand-side uncertainties, so as to
improve the system reliability. We also discuss several advanced solution approaches
and algorithms for specific stochastic mixed integer linear programming models.

3.2   Two-Stage Stochastic Unit Commitment Modeling

Stochastic optimization approach is to apply stochastic programming to model decisions under uncertainties. Here, an important feature is that uncertainties are assumed
to be known and can be presented in a scenario tree. Theoretically, the more scenarios
are involved in a scenario tree, the more comprehensive uncertainties are involved
in that all possible scenarios are assumed to be discrete and independent. We define
the symbol ξ as a scenario in the UC problem. The abstract form of stochastic unit
commitment (SUC) problem can be expressed as follow.

$$[\mathbf{P}] : \min \ \mathbf{c}_1^\mathbf{T}\mathbf{x} + \mathbb{E}((\mathbf{c}_2^\mathbf{T})^\xi \mathbf{y}^\xi) \tag{3.1}$$

$$\text{s.t.} \ \mathbf{A}_1\mathbf{x} = \mathbf{b}_1 \tag{3.2}$$

$$\mathbf{A}_2^\xi\mathbf{x} + \mathbf{E}^\xi\mathbf{y}^\xi = \mathbf{b}_2{}^\xi, \quad \forall \xi \in \varXi \tag{3.3}$$

$$\mathbf{x} \in \{0, 1\}^{n_1} \tag{3.4}$$

$$\mathbf{y}^\xi \in \mathbb{R}_+^{n_2}, \quad \forall \xi \in \varXi \tag{3.5}$$

where $\mathbf{c}_1 \in \mathbb{R}^{n_1}, \mathbf{c}_2 \in \mathbb{R}^{n_2}, \mathbf{b}_1 \in \mathbb{R}^{m_1}, \mathbf{b}_2 \in \mathbb{R}^{m_2}, \mathbf{A}_i \in \mathbb{R}^{n_1 \times m_i} (i = 1, 2), \mathbf{E} \in \mathbb{R}^{n_2 \times m_2}$, and m_1, m_2 are scalars. From the above SUC model, decision variables can be separated to here-and-now variables (i.e. first-stage variables) and wait-and-see variables (i.e. second-stage variables). On a day-ahead power market, the here-and-now decisions are made one day ahead before all uncertainties are revealed in the next day. These here-and-now decisions can directly or indirectly affect wait-and-see decisions, but should ensure sufficient generation resources to deal with forecasted uncertainties on next day.

For the relationship between stochastic unit commitment and deterministic unit commitment, a stochastic UC model with a single scenario can be considered as a deterministic model. In doing so, solving a stochastic UC model is equivalent to solving a large-scale deterministic UC model, but the solution process becomes challengeable.

As we mention above, some common uncertainties that can be described in discrete scenarios include

- forecasted demand $D_{it}^{\xi 0}$,
- renewable energy output R_{it}^{ξ},
- electricity price Q_{it}^{ξ},
- generating unit outage $\alpha_{it}^{\xi} P_{gt}$, and
- transmission element outage, e.g. $\alpha_{ijt}^{\xi} F_{ijt}$.

The first three uncertainty resources usually have successive fluctuations during a certain time period, while the last two uncertainty resources are intermittent occurrences. In stochastic optimization, we can simulate a continuous uncertainty to be a serious of possible random discrete values, which can construct a finite set. Thus, all these possible values as parameters/data inputs can be easily incorporated to SUC models.

From the view of power balance, any changes from uncertainty resources may lead to net load changes on generation and transmission. As dependent variables, the decisions related to above uncertainties are modeled as higher dimensional variables based on each scenario ξ. The following list contains main decisions made in each scenario including, but not limited to:

- dispatch level, p_{gt}^{ξ}
- spinning reserve, s_{gt}^{ξ}
- electric power flow from bus i to bus j, f_{ijt}^{ξ}
- load-shedding loss, δ_{it}^{ξ}
- phase angle, β_{it}^{ξ}
- shifted demand, y_{it}^{ξ}
- energy storage level, r_{it}^{ξ}
- energy storage injection, v_{it}^{ξ}
- energy storage dispatch level, x_{it}^{ξ}.

3.2.1 Problem Formulation

With multiple scenarios involved in the optimization problem, we modify the deterministic objective function (2.1) to the stochastic objective function (3.6) considering the scenario-based representation of variables. The difference is on the second-stage costs which are replaced with the expected fuel costs and penalty costs under all scenarios. Here, let $Prob^\xi$ denote a probability of scenario, usually called a weight for a scenario. The objective function for a stochastic UC model can be written with the second-stage operational costs on average,

$$
\min \sum_{g \in G} \sum_{t \in T} (SU_g v_{gt} + SD_g w_{gt})
$$

$$
+ \sum_{\xi \in \Xi} Prob^\xi \left(\sum_{g \in G} \sum_{t \in T} \left[F_g(p_{gt}^\xi) + F_g(s_{gt}^\xi) \right] + VOLL \sum_{t \in T} \sum_{i \in N} \Delta_{it}^\xi \right) \quad (3.6)
$$

Note the first term of the objective function is the first-stage startup and shut down costs (determined by here-and-now decisions), while the second term associated with weighted scenarios is the expected fuel costs for electricity dispatch and the unserved energy costs.

In the stochastic environment, we may know or estimate the second-stage input data following a known probability distribution (e.g. Normal distribution). Correspondingly, it can provide a discrete scenario with a certain probability. However, in some situations such as generator outage, transmission line outage, demand dramatic changes, renewable energy volatility and intermittence and other contingencies, the probabilities for such scenarios are usually unknown or hard to estimate a probability distribution. One can use stochastic outputs of a simulation model or an empirical distribution from historical data as an approximation for second-stage input data. Therefore, whichever distribution of input data in the second stage is a weighted average of operational costs for all scenarios.

The UC decisions as we know are determined in day ahead (called first stage decisions), more specially before uncertainty realizations. Actually, they are not formulated as scenario-based variables. Beyond these UC decisions, other decision variables affected by uncertainties are considered as scenario-based variables, and their corresponding constraints are located in the second stage of stochastic UC problems. We select some typical scenario-based constraints to show below. For the generator-related constraints,

$$
\left.
\begin{aligned}
&P_g^{min} u_{gt} \leq p_{gt}^\xi \leq P_g^{max} u_{gt}, \\
&p_{gt}^\xi - p_{gt-1}^\xi \leq P_g^{min}(2 - u_{gt} - u_{g(t-1)}) + RU_g(1 + u_{g(t-1)} - u_{gt}), \\
&p_{gt-1}^\xi - p_{gt}^\xi \leq P_g^{min}(2 - u_{gt} - u_{g(t-1)}) + RD_g(1 - u_{g(t-1)} + u_{gt}), \\
&p_{gt} + s_{gt} \leq P_g^{cap} u_{gt}, \\
&s_{gt}^\xi \leq S_g^{max}, \\
&p_{gt}^\xi, \ s_{gt}^\xi \geq 0,
\end{aligned}
\right\}
\begin{aligned}
&\forall g \in G, \ t \in T \\
&\xi \in \Xi
\end{aligned}
$$

For the bus-related constraints,

$$\left.\begin{array}{l} \sum_{g \in G_i} s_{gt}^{\xi} \geq RS_{it}^{\xi}, \\ y_{it}^{\xi} = D_{it}^{\xi 0} + E_{it}^{\xi}(q_{it}^{\xi} - Q_{it}^{\xi 0}), \\ \alpha Q_{it}^{\xi 0} \leq q_{it}^{\xi} \leq \gamma Q_{it}^{\xi 0}, \\ r_{it}^{\xi} = r_{it-1}^{\xi} + v_{it-1}^{\xi} - x_{it-1}^{\xi}, \\ x_{it}^{\xi} \leq r_{it}^{\xi}, \\ r_{it}^{\xi} \leq \kappa_i, \\ \sum_{(i,j) \in A_i^+} f_{ijt}^{\xi} - \sum_{(j,i) \in A_i^-} f_{jit}^{\xi} = \sum_{g \in G_i} p_{gt}^{\xi} + R_{it}^{\xi} - D_{it}^{\xi 0} + \delta_{it}^{\xi}, \\ v_{it}^{\xi}, x_{it}^{\xi}, r_{it}^{\xi}, \delta_{it}^{\xi} \geq 0, \end{array}\right\} \begin{array}{l} \forall \, i \in N, \, t \in T \\ \xi \in \varXi \end{array}$$

For the line-related constraints,

$$\left.\begin{array}{l} (f_{ijt}^{\xi} - f_{jit}^{\xi}) - B_{ijt}^{\xi}(\beta_{it}^{\xi} - \beta_{jt}^{\xi}) = 0, \\ -F_{ij}^{max}\alpha_{ijt} \leq f_{ijt}^{\xi} \leq F_{ij}^{max}\alpha_{ijt}, \\ f_{ijt}^{\xi}, \beta_{it}^{\xi}, \text{ unrestricted} \end{array}\right\} \forall \, (i, j) \in A, \, i \in N, \, t \in T, \, \xi \in \varXi$$

Here we continue using discrete scenarios within the scenario set $\varXi$ to model supply and demand side uncertainties. In practice, simulation technique is very helpful to generate random numbers through a simulation model and simulate real-time scenarios as an optimization model's data input. As the second stage is to address the next-day decisions (i.e. power dispatch, ramping, reserve, power flow, etc.) for all scenarios, their features are captured by $|\varXi|$ sets of continuous variables and constraints. Meanwhile, because the unit commitment decisions will affect next-day decisions, dispatch constraints in the second stage work in connection with day-ahead UC decisions.

The second stage of SUC model is mainly to solve next-day operational problems with respect to power generation/dispatch, power transmission, and other available resources after known uncertainties unfold. What's more, an optimal operation schedule must satisfy the criteria of system's reliability, and thus some research implemented VaR or CVaR-based risk constraints in the second stage, instead of performing contingency feasibility check. As an example, the integrated two-stage SUC model **SUCR-DR-ES** includes UC constraints (3.7)–(3.15), demand response constraints (3.16)–(3.17), energy storage constraints (3.18)–(3.20), power transmission constraints (3.22)–(3.24), and CVaR risk constraints (3.25)–(3.26), stated as follow.

$$\min \sum_{g \in G} \sum_{t \in T} (SU_{gt}v_{gt} + SD_{gt}w_{gt}) + \sum_{\xi \in \varXi} Prob^{\xi} \sum_{t \in T} \sum_{g \in G} [(b_{gt}p_{gt}^{\xi} + a_{gt}u_{gt})]$$

$$\text{s.t. } u_{gt} - u_{g(t-1)} \leq u_{g\tau}, \quad \forall \, g \in G, \, t \in T, \tau = t, \dots, min\{t + L_g - 1\} \quad (3.7)$$

$$u_{g(t-1)} - u_{gt} \leq 1 - u_{g\tau}, \quad \forall \, g \in G, \, t \in T, \tau = t, \dots, min\{t + l_g - 1\} \quad (3.8)$$

$$v_{gt} \geq u_{gt} - u_{g(t-1)}, \quad \forall \, g \in G, \, t \in T \quad (3.9)$$

$$w_{gt} \geq -u_{gt} + u_{g(t-1)}, \quad \forall g \in G, \, t \in T \tag{3.10}$$

$$u_{gt}, \, v_{gt}, \, w_{gt} \in \{0, 1\}, \quad \forall g \in G, \, t \in T \tag{3.11}$$

$$P_g^{min} u_{gt} \leq p_{gt}^{\xi} \leq P_g^{max} u_{gt}, \quad \forall g \in G, \, t \in T, \, \xi \in \Xi \tag{3.12}$$

$$- RD_g \leq p_{gt}^{\xi} - p_{gt-1}^{\xi} \leq RU_g, \quad \forall g \in G, \, t \in T, \, \xi \in \Xi \quad \leftarrow \chi_{gt}^{\xi}, \, \sigma_{gt}^{\xi} \tag{3.13}$$

$$s_{gt}^{\xi} \leq S_g^{max}, \quad \forall g \in G, t \in T, \, \xi \in \Xi \quad \leftarrow \upsilon_{gt}^{\xi} \tag{3.14}$$

$$\sum_{g \in G_i} s_{gt}^{\xi} \geq RS_{it}, \quad \forall i \in N, t \in T, \, \xi \in \Xi \quad \leftarrow \lambda_{it}^{\xi} \tag{3.15}$$

$$y_i^{\xi} = D_i^0 + E_{it}^{\xi}(q_{it}^{\xi} - Q_{it}^{\xi}), \quad \forall i \in N, t \in T, \, \xi \in \Xi \quad \leftarrow \mu_{it}^{\xi}, \, \nu_{it}^{\xi} \tag{3.16}$$

$$\alpha Q_{it}^{\xi} \leq q_{it}^{\xi} \leq \gamma Q_{it}^{\xi}, \quad \forall i \in N, t \in T, \, \xi \in \Xi \tag{3.17}$$

$$r_{it}^{\xi} = r_{it-1}^{\xi} + v_{it-1}^{\xi} - x_{it-1}^{\xi}, \quad \forall i \in N, \, t \in T, \, \xi \in \Xi \tag{3.18}$$

$$0 \leq x_{it}^{\xi} \leq r_{it}^{\xi}, \quad \forall i \in N, \, t \in T, \, \xi \in \Xi \tag{3.19}$$

$$0 \leq r_{it}^{\xi} \leq \kappa_i, \quad \forall i \in N, \, t \in T, \, \xi \in \Xi \quad \leftarrow \vartheta_{it}^{\xi} \tag{3.20}$$

$$\sum_{(i,j) \in A_i^+} f_{ijt}^{\xi} - \sum_{(j,i) \in A_i^-} f_{jit}^{\xi} - \sum_{g \in G_i} p_{gt}^{\xi} - \Delta_{it}^{\xi} + v_{it}^{\xi} - \rho_i x_{it}^{\xi} = W_{it}^{\xi} - y_{it}^{\xi}, \tag{3.21}$$

$$\forall i \in N, \, t \in T, \, \xi \in \Xi \tag{3.22}$$

$$(f_{ijt}^{\xi} - f_{jit}^{\xi}) - B_{ijt}(\beta_{it}^{\xi} - \beta_{jt}^{\xi}) = 0, \quad \forall \, (i, j) \in A, \, t \in T, \, \xi \in \Xi \tag{3.23}$$

$$- F_{ij}^{max} \alpha_{ijt} \leq f_{ijt}^{\xi} \leq F_{ij}^{max} \alpha_{ijt}, \quad \forall \, (i, j) \in A, \, t \in T, \, \xi \in \Xi \tag{3.24}$$

$$\sum_{i \in I} \Delta_{it}^{\xi} \leq \eta_t + \zeta_t^{\xi}, \quad \forall \, t \in T, \, \xi \in \Xi \tag{3.25}$$

$$\eta_t + (1 - \theta)^{-1} \sum_{\xi \in \Xi} \text{Prob}^{\xi} \zeta_t^{\xi} \leq \bar{\phi}, \quad \forall \, t \in T \tag{3.26}$$

$$\eta_t \geq 0, \, \zeta_t^{\xi} \geq 0, \quad \forall \, t \in T, \, \xi \in \Xi \tag{3.27}$$

$$p_{gt}^{\xi}, \, s_{gt}^{\xi} \geq 0, \quad \forall \, g \in G, \, t \in T, \, \xi \in \Xi \tag{3.28}$$

$$\Delta_{it}^{\xi}, \, v_{it}^{\xi} \geq 0, \quad \forall \, i \in N, t \in T, \, \xi \in \Xi \tag{3.29}$$

$$f_{ijt}^{\xi}, \quad \forall \, (i, j) \in \mathscr{A}, t \in T \tag{3.30}$$

3.3 SUC with Demand Respond

Electricity demand has a lot of historical data that can be accessed to provide useful information in demand forecast. However, it is considered as one of uncertain sources in a power system because it may have significant changes at any moment.

As demand sides are connected to supply sides through transmission network, any demand change nearly affects the electricity supply instantly and requires an immediate response of power generation. This response requirement can be supported by monitoring electricity consumptions on each bus. To further control electric consumptions, ISOs/RTOs promote demand response programs so as to adjust electricity demand based on current power system's operations and generation capacities.

3.3.1 Demand Respond

As one of common non-generation resources, demand response (DR) has been implemented and developed for several years in some energy markets, e.g. PJM, NYISO, CAISO. Demand Response program aims to encourage end users to lower power consumption during peak hours but increase consumption during off-peak hours or high production times. In practice, demand side management has shown as an effective tool to mitigate loads at peak hours (technically measured by peak-to-average ratio), which is generally executed through demand response programs.

Demand change is a reflection of consumers' responding behaviors. In addition to daily basic demands, there are many factors that can cause demand changes, such as electricity price, weather conditions, household renewable energy outputs and social events.

To avoid over use of costly generators (e.g. quick-start generators), some operators investigate the relationships between responsive demands and price signals, which can be modeled as a fixed price–elastic demand curve. So they can adjust electricity prices to control demands within a reasonable range. The price elasticity is defined as the percentage change of demand quantity, with the consideration of a small change in price. Figure 3.1 shows the price–elastic demand curve that reveals the essential relationship between responsive demands and locational marginal prices. The supply curve and the demand curve are shown in red curve and black curve respectively, and the intersection on these two curves indicates a demand d_{it} on price p_{it}.

For power system's operations, the major uncertainty from demand side comes from the responding behaviors of end users on varying electricity prices. For the simplicity, the real-time demand is assumed to consist of the forecasted inelastic demands and the demand changes due to electricity prices Q_i^ξ (presented by price elasticity matrix $\mathbf{E}_i^\xi$). Because renewable energy is viewed as an uncertainty and has a relationship to DR in models, we here suppose that the outputs of any renewable energy is independent with both DR and locational price at a bus within a planning horizon. In doing so, small demand change subject to varying electricity price can be described through a price elasticity matrix.

In fact, if price variation is small as in [WKK10], price–elastic demand can be formulated by a set of approximate linear constraints associated with a price elasticity matrix. Constraint (3.31) states that the forecasted real-time demand is the summation of benchmark demand $\mathbf{D}_i^0$ and elastic demand caused by the price difference between real-time price and benchmark price, $\mathbf{E}_i(\mathbf{q}_i - \mathbf{Q}_i^0)$.

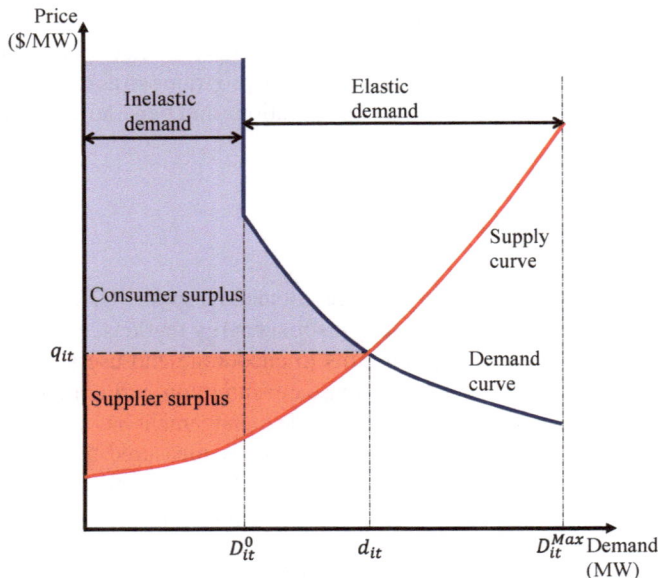

Fig. 3.1 Relationship between demands and prices [ZWWG13]

$$\mathbf{y}_i = \mathbf{D}_i^0 + \mathbf{E}_i(\mathbf{q}_i - \mathbf{Q}_i^0) \qquad\qquad \forall\, i \in N \qquad\qquad (3.31)$$

In additions, electricity price constraint (3.32) is help to maintain real-time price fluctuation within an expected price range.

$$\alpha \mathbf{Q}_i^0 \le \mathbf{q}_i \le \gamma \mathbf{Q}_i^0 \qquad\qquad \forall\, i \in N \qquad\qquad (3.32)$$

where
    $\mathbf{y}_i$:   shifted demand at bus i for all time periods
    $\mathbf{q}_i$:   real-time electricity price at bus$i$
    $\mathbf{D}_i^0$:   benchmark/reference demand vector at bus i for all time periods
    $\mathbf{E}_i^\xi$:   the price elasticity matrix
    $\mathbf{Q}_i^0$:   benchmark/reference electricity price vector at bus i for all time periods
    α, γ : price velocity indicators

In the above linear constraints, both $\mathbf{y}_i$ and $\mathbf{q}_i$ are continuous decision variables shown in vector. Each of the vectors contain the decision variables of different time periods, e.g., $\mathbf{y}_i = [y_{it}, \ \forall t \in T]^T$.

In above linear constraints, we note that a shifted demand at time t, y_{it}, is an affine function regard to price variations within a time period, and dependent on the benchmark demand (day-ahead forecast) $\mathbf{D}_i^0$ at time t. It is reasonable to assume that the total demand of all time periods is a constant, especially formulated under a certain scenario. Through the price elasticity matrix $\mathbf{E}_i$ for each bus, we can easily manage price variations that are reflected in demand variations.

Since the function of DR is to adjust power balance on demand sides. When it connects to a transmission network to fulfill this function, the results from real-time demand adjustment eventually impact thermal generation in the same power systems. The KCL transmission formulation is shown.

$$\sum_{(i,j)\in A_i^+} f_{ijt} - \sum_{(j,i)\in A_i^-} f_{jit} = \sum_{g\in G_i}(p_{gt} + s_{gt}) + R_{it} + \delta_{it} - y_{it}, \quad \forall i \in N, \ t \in T$$

(3.33)

With the participation of DR, the forecasted demands D_{it}^0 become the locally adjusted demands y_{it} influenced by demand response program.

To mitigate forecast errors for day-ahead renewable energy resource, DR programs is also viewed as another reserve resource in addition to traditional reserve. The UC integrated with DR program and wind-based generation has been fully discussed in [Sio10, ZZ12, DLOR12, KM11]. The price–elastic demand curve is proposed as another different method to express the relationship between electricity price and demand. Actually any price–elastic demand curve is hardly known exactly in advance, thus it still causes uncertainty for UC scheduling. What's more, the renewable energy output varying within a given interval may induce the change of price–elastic demand curve. Assuming that the response of supply changes is instantaneous, there is an existing situation that thermal generation output would be reduced as wind energy output increased. Considering a load area with single electricity price, the total supply from thermal and renewable can meet local demand, but the real-time locational marginal price is subject to be reduced, probably leading to demand increase.

Figure 3.2 shows the relationship between demand and price, where uncertainty of price–elastic demand curve is allowed to occur within a specific range (grey area). Based on the price–elastic demand curve, when an electricity price q_{it}^0 is given, the corresponding demand d_{it} remains uncertain and could be any possible value between upper limit and lower limit (The range for d_{it} within two dashed line). Conversely, if a demand is fixed at time t, there exists a uncertain price q_{it} with a corresponding price range. The formulation of price–elastic demand curve is

$$d_{it} = A_{it}(q_{it})^\alpha + \varepsilon_{it}$$

or

$$d_{it} = A_{it}(q_{it} + \varepsilon_{it})^\alpha,$$

where ε_{it} is defined as a deviation from the original price Q_{it}^0 or generally said the uncertainty of price–elastic demand curve.

To linearize the price–elastic demand curve, it can be approximated as a step-wise curve against each demand d_{it}. Thus, the corresponding forecasted price q_{it}^k varies

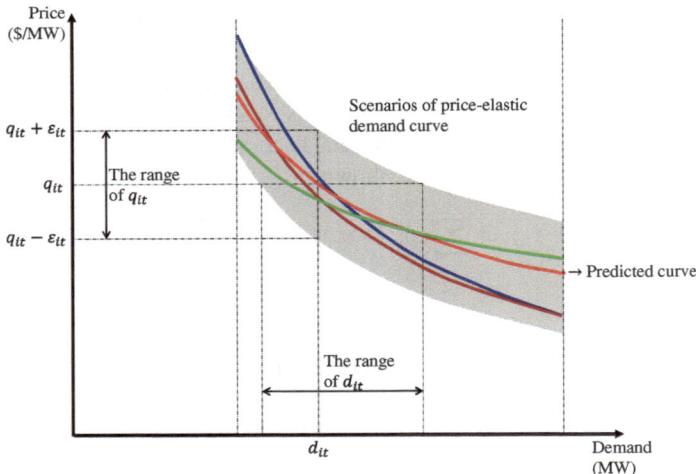

Fig. 3.2 Uncertainty of price–elastic demand curve [ZWWG13]

within the range $q_{it}^k = [q_{it}^{k'} - \hat{\varepsilon}_{it}^k, q_{it}^{k'} + \hat{\varepsilon}_{it}^k]$, where $q_{it}^{k'}$ is the forecasted value of q_{it}^k and $\hat{\varepsilon}_{it}$ is the maximum deviation of ε_{it}^k. Moreover, to control the span of deviation ε_{it}^k on each step, it is necessary to apply a hard limit for restricting the total amount of deviations $\sum_{k \in K} \varepsilon_{it}^k$. The uncertainty set of DR curve is shown as follow:

$$\left\{ \varepsilon : -\hat{\varepsilon}_{it}^k \leq \varepsilon_{it}^k \leq \hat{\varepsilon}_{it}^k, \ -\varpi_{it} \leq \sum_{k \in K} \varepsilon_{it}^k \leq -\varpi_{it}, \ \forall \, t \in T, i \in N, k \in K \right\}.$$

3.4 SUC with Energy Storage

As one of typical non-generation resources, energy storage (ES) is an complementary approach to capture produced energy when the generation exceed demands and provide power supply later when the peak demand is tending to over the power generation capacities, or the generation costs are extremely high.

Currently, any electrical power system has to adopt power generation to power consumption immediately according to the physical law on power circuits. This commitment makes a power systems encounter many issues, e.g., insufficient generation or transmission line capacities, reserve redundancy, or supply and demand imbalance, *etc*. Taking the ES's benefits, these issues can be solved or mitigated through the appropriate uses of ES technologies in a power system, i.e., pumped hydroelectric storage, compressed air energy storage, and different types of rechargeable batteries (battery banks).

As we know the uncertainties of renewable energy generation, ES works like a buffer constantly saving and providing power at a moment. The intermittent and uncertain nature of renewable energy outputs can be weakened by the operations on energy storage devices.Thus, we are putting a lot of effort to deploy energy storage and expecting it to provide a great assistance to expand the use of renewable energy. The large deployment of energy storage relies on the seamless integration of energy storage operations with existing power system operations. This is an essential step to ensure power system's reliability and flexibility, resolve issues of renewable energy penetration, and even promote deployment of electric transportation [U.S13].

3.4.1 Energy Storage

With respect to operation and production planning, Rong et al. [RLL08] presents a deflected subgradient optimization method applied for trigeneration planning with storages. The model involves the charge/discharge and physical conditions of energy storage. To evaluate an optimal scheduling of energy storage system, one also should consider the dependence of storage sizing, the cost of energy delivery on renewable energy levels as well as storage efficiency [AJ09]. Chakraborty et al. [CTT+09] study the optimal size of energy storage systems also impacts the further UC scheduling of power systems. DeJonghe et al. [JDBD11] address how energy storage can better facilitate the power technology mix with wind power penetration by increasing base load capacity.

Considering the operations of ES in stochastic environments, there are a group of energy storage constraints to present the power status for energy storage accumulators and the power charge/discharge in a ES storage system level at each time period. Constraint (3.34) reflects the current power in a battery bank after charging or discharge operation in the last period, where charging action and discharging action would not occur simultaneously because the power loss due to discharge from a battery is considered in the KCL transmission constraint.

$$r_{it} = r_{it-1} + v_{it-1} - x_{it-1} \qquad \forall i \in N, t \in T \qquad (3.34)$$
$$v_{it} \geq 0 \qquad \forall i \in N, t \in T \qquad (3.35)$$

The other constraints (3.36) and (3.37) separately show a power discharge from battery bank restricted by an available power in battery bank and a power storage capacity.

$$0 \leq x_{it} \leq r_{it} \qquad \forall i \in N, t \in T \qquad (3.36)$$
$$0 \leq r_{it} \leq \kappa_i \qquad \forall i \in N, t \in T \qquad (3.37)$$

where

r_{it} :   the total remaining power in storage facilities of unit i at time t
v_{it-1} :   the power storage at bus i in period t
x_{it-1} :   the power discharge amount at bus i in period t
κ_i :   the maximum storage capacity at bus i

Note that N can be replaced by a subset $N' \subset N$, since ES bank may not be available at each bus.

In most cases, ES is mainly considered to be on the power supply side to adjust power balance. When it is connected to a transmission network, power charge and discharge would affect thermal generation in the same power system. If a power system is without any demand-side management, the forecasted demand is normally treated as a benchmark demand $D_{it}^{\xi 0}$. In fact, the energy storage has two actions, i.e. consuming power from a bus (the hourly amount of power saving denoted as **v**) and supplying power to the electric grid (the hourly amount of power dispatch denoted as **x**). The KCL transmission constraint is rewritten including the ES operations, as follow.

$$\sum_{(i,j)\in A_i^+} f_{ijt} - \sum_{(j,i)\in A_i^-} f_{jit} = \sum_{g\in G_i}(p_{gt} + s_{gt}) + R_{it} + \rho_i x_{it} + \delta_{it} - v_{it} - D_{it}^0,$$

$$\forall i \in N,\, t \in T \tag{3.38}$$

where ρ_i denotes the ES efficiency mainly depending on storage device properties.

When both ES and DR programs are implemented at some nodes, we can revise the KCL transmission constraint to accommodate the process of power saving and dispatch simultaneously with adjusted demands. There are many studies to address the cost saving effects about the implementation of individual DR and ES programs, but the combined two resources are superior and capable of reducing the total expected operational costs for their joint actions. The following KCL constraint is suitable for this situation,

$$\sum_{(i,j)\in A_i^+} f_{ijt} - \sum_{(j,i)\in A_i^-} f_{jit} = \sum_{g\in G_i}(p_{gt} + s_{gt}) + R_{it} + \rho_i x_{it} + \delta_{it} - v_{it} - y_{it},$$

$$\forall i \in N,\, t \in T. \tag{3.39}$$

3.4.2 Case 3: Two-Stage Stochastic Unit Commitment with Energy Storage and Wind Power Generation

This case focuses on the optimization of UC with a non-generation resource under a series of wind energy scenarios. To ensure the system security, the spinning reserve

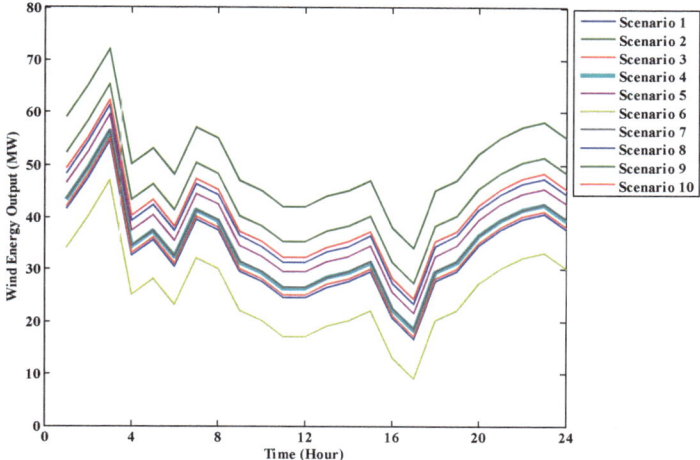

Fig. 3.3 Case 2: Hourly wind energy outputs for 10 scenarios

service remains in stochastic UC model. This case is modeled as a two-stage stochastic mixed-integer linear program to adopt stochastic wind energy outputs.

Energy storage is selected as non-generator resource to offer energy supply and
storage. Only Bus 1, 2, 4 and 5 are qualified to utilize energy storage devices which
have their own storage capacities (see Fig. 2.1). The ES efficiency ρ_i is assumed to
be 0.95 for all devices. In this case, any costs for energy storage are not taken into
account.

Wind energy output is one of main determining stochastic factors affecting power
generation levels. For the simplicity of stochastic environment, only wind energy
output is considered as a typical renewable energy resource and is simulated through
base wind energy output patterns. Although wind speed distribution generally can be
described by Weibull distribution, the wind energy output is assumed to be normally
distributed for easy access. All scenarios can be produced by obtaining random
numbers through C++ normal distribution generators and aggregating them with
the hourly base load. We provide an scenario generation example in Appendix B,
in which the wind scenarios' data are generated as model input parameters and
presented in Fig. 3.3. It is assumed that the probability for each scenario is known,
$prob^{\xi} = 10\%, \forall \xi \in \Xi$.

The index sets are given below:

$G = 4$ Generators
$T = 24$ Hours
$N = 7$ Buses
$S = 10$ Scenarios
$|\mathscr{A}| = 10$ Transmission Lines

The abstract form of stochastic UC is presented as follow.

$$\min \sum_{g \in G} \sum_{t \in T} (SU_{gt}v_{gt} + SD_{gt}w_{gt}) + \sum_{\xi \in \Xi} Prob^{\xi} \sum_{t \in T} \sum_{g \in G} [(b_{gt}p_{gt}^{\xi} + a_{gt}u_{gt})$$

$$+ (b'_{gt}s_{gt}^{\xi} + a'_{gt}u_{gt})] + \sum_{\xi \in \Xi} Prob^{\xi} VOLL \sum_{t \in T} \sum_{i \in N} \Delta_{it}^{\xi}$$

s.t. The first-stage constraints

$$u_{gt} - u_{g(t-1)} \leq u_{g\tau}, \quad \forall g \in G, t \in T, \tau = t, \ldots, min\{t + L_g - 1, |T|\}$$

$$u_{g(t-1)} - u_{gt} \leq 1 - u_{g\tau}, \quad \forall g \in G, t \in T, \tau = t, \ldots, min\{t + l_g - 1, |T|\}$$

$$v_{gt} \geq u_{gt} - u_{g(t-1)}, \quad \forall g \in G, t \in T$$

$$w_{gt} \geq -u_{gt} + u_{g(t-1)}, \quad \forall g \in G, t \in T$$

$$u_{gt}, v_{gt}, w_{gt} \in \{0, 1\}, \quad \forall g \in G, t \in T$$

The second-stage constraints

$$P_g^{min}u_{gt} \leq p_{gt}^{\xi} \leq P_g^{max}u_{gt}, \quad \forall g \in G, t \in T, \xi \in \Xi$$

$$- RD_g \leq p_{gt}^{\xi} - p_{gt-1}^{\xi} \leq RU_g, \quad \forall g \in G, t \in T, \xi \in \Xi$$

$$0 \leq s_{gt}^{\xi} \leq S_g^{max}, \quad \forall g \in G, t \in T, \xi \in \Xi$$

$$p_{gt}^{\xi} + s_{gt}^{\xi} \leq P_g^{cap}u_{gt}, \quad \forall g \in G, t \in T, \xi \in \Xi$$

$$\sum_{g \in G_i} s_{gt}^{\xi} \geq RS_{it}, \quad \forall i \in N, t \in T, \xi \in \Xi$$

$$r_{it}^{\xi} = r_{it-1}^{\xi} + v_{it-1}^{\xi} - x_{it-1}^{\xi}, \forall i \in N, t \in T, \xi \in \Xi$$

$$0 \leq x_{it}^{\xi} \leq r_{it}^{\xi}, \forall i \in N, t \in T, \xi \in \Xi$$

$$0 \leq r_{it}^{\xi} \leq \kappa_i, \forall i \in N, t \in T, \xi \in \Xi$$

$$\sum_{(i,j) \in A_i^+} f_{ijt}^{\xi} - \sum_{(j,i) \in A_i^-} f_{jit}^{\xi} - \sum_{g \in G_i} (p_{gt}^{\xi} + s_{gt}^{\xi}) - \Delta_{it}^{\xi} + v_{it}^{\xi} - x_{it}^{\xi} = W_{it}^{\xi} - D_{it},$$

$$\forall i \in N, t \in T, \xi \in \Xi$$

$$(f_{ijt}^{\xi} - f_{jit}^{\xi}) - B_{ijt}(\beta_{it}^{\xi} - \beta_{jt}^{\xi}) = 0, \quad \forall (i, j) \in A, t \in T, \xi \in \Xi$$

$$- F_{ij}^{max}\alpha_{ijt} \leq f_{ijt}^{\xi} \leq F_{ij}^{max}\alpha_{ijt}, \quad \forall (i, j) \in A, t \in T, \xi \in \Xi$$

$$p_{gt}^{\xi}, s_{gt}^{\xi} \geq 0, \quad \forall g \in G, t \in T, \xi \in \Xi$$

$$\Delta_{it}^{\xi}, v_{it}^{\xi} \geq 0, \quad \forall i \in N, t \in T, \xi \in \Xi$$

$$f_{ijt}^{\xi}, \quad \forall (i, j) \in \mathscr{A}, t \in T$$

If readers are interested in detailed stochastic unit commitment modeling, one can refer to Appendix A for explicit definitions of symbols.

Table 3.1 Objective value and stochastic UC including ES for 7-bus system

Objective value	Unit ID	Hour (1–24)
$60615.6	G1	1 1 1 1 1 1 1 1 1 1 1 1 1 1 1 1 1 1 1 1 1 1 1 1
	G2	1 1 1 1 1 1 1 1 1 1 1 1 1 1 1 1 1 1 1 1 1 1 1 1
	G3	0 0 0 0 0 0 0 0 0 0 0 0 0 0 0 0 0 1 1 1 1 0 0 0 0
	G4	0 0 0 0 0 0 0 0 0 1 1 1 1 1 1 1 1 1 1 1 1 1 1 1

Table 3.2 Objective value and stochastic UC excluding ES for 7-bus system

Objective value	Unit ID	Hour (1–24)
$65274.8	G1	1 1 1 1 1 1 1 1 1 1 1 1 1 1 1 1 1 1 1 1 1 1 1 1
	G2	1 1 1 1 1 1 1 1 1 1 1 1 1 1 1 1 1 1 1 1 1 1 1 1
	G3	0 0 0 0 0 0 0 0 0 0 0 0 0 0 0 0 1 1 1 1 1 0 0 0 0
	G4	0 0 0 0 0 0 1 1 1 1 1 1 1 1 1 1 1 1 1 1 1 1 1 1

Solving a stochastic UC model is equivalent to solving multiple deterministic models with given scenario probabilities. This model is solved using C++/CPLEX, where the Brand-and-Cut-and-Price algorithm is applied to solve this type of stochastic mixed integer linear program. See Sect. 3.5 for more detailed solution techniques and advanced solution approaches.

Tables 3.1 and 3.2 show the objective value and the stochastic UC schedules including ES resource and excluding ES resource, respectively. Compared to their objective values, the UC with ES has a lower total generation cost than that of UC without ES. The energy storage capacities hold around 15% of generation capacities and successfully help reducing 7.18% of daily generation costs based on the basic power system (without ES). In addition, the UC schedule on Table 3.1 shows less commitments needed since ES is able to flexibly store electricity in off-peak hours and offer additional electricity in on-peak hours. In doing so, ES helps not only releasing limited generation resources, but also maintaining system reliability.

We also compare the results from deterministic model (Table 2.4) and stochastic model (Table 3.2). The difference between these two models is the number of wind energy output scenario, where 60% of scenarios have lower wind outputs than scenario 1. As more scenarios involved, the thermal generation system has to put more generators online to handle the uncertainty of low wind energy outputs. It can be found that G4 in the stochastic case is assigned with online task starting from 7am until midnight. The G3 online hours also are extended forward one more hour to satisfy the peak load. Meanwhile, the high probability of low wind output naturally lead to higher expected thermal generation costs.

3.5 Two-Stage Stochastic Unit Commitment Problem Decomposition

For the cases with large bus systems, solving an original SUC model directly by a solver is time consuming. We can adopt a decomposition approach to shorten the computation time for obtaining optimal solutions. A decomposition approach generally break an original model into an integer program of **RMP** and one or more linear programs for the subproblem. As an example of the classical Benders' decomposition strategy, a two-stage SUC model will be intuitively decomposed into the **RMP** only with UC constraints and the subproblem with next-day operation constraints.

In the fact that this decomposition may bring in low-density cuts, such as a cut only including a single decision variable **u**, the solving process will converge slowly. Also, if a subproblem has a coupling constraint, i.e. all scenarios coupled by risk constraints, this may potentially restrict applying parallel computation resources during the solving process. Thus, an alternative decomposition strategy was proposed to resolve this issue aiming to increase the density of Benders cuts and mitigate the impacts from coupling constraints.

As a coupling constraint shows in the CVaR constraints (3.26), we can move all CVaR constraints to **RMP**, and this allows incumbent solutions $(\mathbf{u}, \mathbf{l})$ involving binary variables and continuous variables that will be passed to $\mathbf{SP}^\xi$. In this way, Benders cuts will include loss variable $\mathbf{l}$ and be able to cut off equivalent or more solution space of **RMP** in an iteration. As for multiple complex subproblems without coupling constraints, they can be solved by parallel computation approach to improve computation time.

We define a free variable π^ξ for the minimum operational cost in a scenario. If the load-shedding loss penalty is not considered in the objective function, the **RMP** with CVaR constraints can be presented by,

$$[\textbf{RMP}] : \min \sum_{g \in G} \sum_{t \in T} (SU_{gt}v_{gt} + SD_{gt}w_{gt}) + \sum_{\xi \in \Xi} \text{Prob}^\xi \pi^\xi$$
$$\text{s.t. } (3.7)-(3.11), \ (3.25)-(3.27),$$
$$\mathbb{F}(u_{gt}, l_{it}^\xi, \pi^\xi) \geq 0, \ \forall \ \xi \in \Xi$$

where constraint $\mathbb{F}(u_{gt}, l_{it}^\xi, \pi^\xi) \geq 0$ is a Benders' cut associated with the commitment variable u_{gt} and loss variable l_{it}^ξ. This cut is generated based on a dual solution from the subproblem for one scenario.

As the subproblem is part of the original **SUCR-DR-ES**, it may not always have optimal solutions to generate optimality cuts. We thus can use the Big-M method in the primal subproblem to avoid getting infeasible solution, but ensure the $\mathbf{SP}^\xi$ maintain the feasibility with incumbent solutions. Recall that the Big-M method will introduce surplus artificial variables which are accordingly penalized in the objective function. We introduce an artificial variable ω_{it} to the system-level spinning reserve constraint (3.15) as well as two positive artificial variables o_{it}^+, and o_{it}^- to the KCL transmission constraint (3.22).

Also, the objective function of $\mathbf{SP}^\xi$ is modified to include the penalty term, a summation of artificial variables associated with a penalty coefficient M. When any of artificial variables is greater than zero, the artificial penalty will occur in the objective function and the $\mathbf{SP}^\xi$ has an optimal solution to produce an optimality Bender's cut. In this case, once we obtain the incumbent solutions $(\mathbf{u}, \mathbf{l})$, we can construct the subproblem (3.40) using the Big-M method for specific second-stage operation constraints.

$$[\mathbf{SP}^\xi] : \min \sum_{g \in G} \sum_{t \in T} F_g(p_{gt}^\xi) + M \sum_{t \in T} \sum_{i \in N} (\omega_{it} + o_{it}^+ + o_{it}^-)$$

$$\text{s.t. } (3.13), (3.14) - (3.20), (3.22) - (3.24) \tag{3.40a}$$

$$p_{gt}^\xi \geq P_g^{min} \hat{u}_{gt}, \quad \forall g \in G, \ t \in T \quad \leftarrow \ \varepsilon_{gt}^\xi \tag{3.40b}$$

$$p_{gt}^\xi + s_{gt}^\xi \leq P_g^{max} \hat{u}_{gt}, \quad \forall g \in G, t \in T \quad \leftarrow \ \rho_{gt}^\xi \tag{3.40c}$$

$$\sum_{g \in G_i} s_{gt}^\xi + \omega_{it} \geq RS_{it}^\xi, \quad \forall i \in N, t \in T \quad \leftarrow \ \tau_{it}^\xi \tag{3.40d}$$

$$\sum_{(i,j) \in A_i^+} f_{ijt}^\xi - \sum_{(j,i) \in A_i^-} f_{jit}^\xi - \left(\sum_{g \in G_i} p_{gt}^\xi + \rho_i x_{it}^\xi - v_{it}^\xi - y_{it}^\xi \right) + o_{it}^+ - o_{it}^-$$

$$= R_{it}^\xi + \hat{l}_{it}^\xi, \quad \forall i \in N, \ t \in T \quad \leftarrow \ \varphi_{it}^\xi \tag{3.40e}$$

where the subproblem usually includes all scenario-independent constraints, including but not limited to, economic power dispatch, ramping, non-generation resources and power transmissions.

3.5.1 Benders' Cut

To build a Benders' cut, we need to define a set of continuous non-negative dual variables to capture the dual values during the solving process. In the **SUCR-DR-ES** model, the dual variables, such as ε_{gt}^ξ, ρ_{gt}^ξ, χ_{gt}^ξ, σ_{gt}^ξ, τ_{it}^ξ, v_{gt}^ξ, λ_{it}^ξ, μ_{it}^ξ, v_{it}^ξ, ϑ_{it}^ξ, φ_{it}^ξ, are assigned to the corresponding constraints (3.40b), (3.40c), (3.13), (3.40d), (3.14), (3.15), (3.16), (3.20), (3.40e). For example, dual variables χ_{gt}^ξ and σ_{gt}^ξ denote the ramping up constraint and the ramping down constraint in (3.13), respectively. After solving the $\mathbf{SP}^\xi$, one can obtain all optimal dual values corresponding to the above constraints. These dual values for one scenario (e.g. $\hat{\chi}_{gt}^\xi$ and $\hat{\sigma}_{gt}^\xi$) are then used to construct an optimality cut $\mathbb{F}(u_{gt}, l_{it}^\xi, \pi^\xi)$, which is presented in (3.41).

$$
\pi^\xi \geq \sum_{g \in G} \sum_{t \in T} \hat{\varepsilon}^\xi_{gt} P^{min}_g u_{gt} + \sum_{g \in G} \sum_{t \in T} \hat{\rho}^\xi_{gt} P^{max}_g u_{gt} + \sum_{g \in G} \sum_{t \in T} \hat{\chi}^\xi_{gt} RU_g
$$

$$
+ \sum_{g \in G} \sum_{t \in T} \hat{\sigma}^\xi_{gt} RD_g + \sum_{i \in N} \sum_{t \in T} \hat{\tau}^\xi_{it} RS^\xi_{it} + \sum_{g \in G} \sum_{t \in T} \hat{\upsilon}^\xi_{gt} S^{max}_g
$$

$$
+ \sum_{i \in N} \sum_{t \in T} \hat{\lambda}^\xi_{it} (D^0_{it} - E^\xi_{it} Q^0_{it}) + \sum_{i \in N} \sum_{t \in T} \hat{\mu}^\xi_{it} \alpha Q^0_{it} + \sum_{i \in N} \sum_{t \in T} \hat{v}^\xi_{it} \gamma Q^0_{it}
$$

$$
+ \sum_{i \in N} \sum_{t \in T} \hat{\vartheta}^\xi_{it} \kappa_i + \sum_{i \in N} \sum_{t \in T} \hat{\varphi}^\xi_{it} (R^\xi_{it} + l^\xi_{it}) \tag{3.41}
$$

Prior to solving **RMP** in the next iteration, the $|\Xi|$ of cuts will be added to **RMP** in theory. In this way, a large number of cuts are accumulated to **RMP**, probably leading to increase the computation times in latter iteration. Alternatively, a single optimality cut for all scenarios $\mathbb{F}(u_{gt}, l^\xi_{it}, \pi^\xi, \forall\, \xi \in \Xi)$ can replace with multiple optimality cuts based on each scenario, which is presented in (3.42).

$$
\pi \geq \sum_{\xi \in \Xi} prob^\xi \left[\sum_{g \in G} \sum_{t \in T} \hat{\varepsilon}^\xi_{gt} P^{min}_g u_{gt} + \sum_{g \in G} \sum_{t \in T} \hat{\rho}^\xi_{gt} P^{max}_g u_{gt} + \sum_{g \in G} \sum_{t \in T} \hat{\chi}^\xi_{gt} RU_g \right.
$$

$$
+ \sum_{g \in G} \sum_{t \in T} \hat{\sigma}^\xi_{gt} RD_g + \sum_{i \in N} \sum_{t \in T} \hat{\tau}^\xi_{it} RS^\xi_{it} + \sum_{g \in G} \sum_{t \in T} \hat{\upsilon}^\xi_{gt} S^{max}_g
$$

$$
+ \sum_{i \in N} \sum_{t \in T} \hat{\lambda}^\xi_{it} (D^0_{it} - E^\xi_{it} Q^0_{it}) + \sum_{i \in N} \sum_{t \in T} \hat{\mu}^\xi_{it} \alpha Q^0_{it} + \sum_{i \in N} \sum_{t \in T} \hat{v}^\xi_{it} \gamma Q^0_{it}
$$

$$
\left. + \sum_{i \in N} \sum_{t \in T} \hat{\vartheta}^\xi_{it} \kappa_i + \sum_{i \in N} \sum_{t \in T} \hat{\varphi}^\xi_{it} (R^\xi_{it} + l^\xi_{it}) \right] \tag{3.42}
$$

3.5.2 The Implementation of Benders' Decomposition

The solving process for classic Bender's decomposition would generate one or more Bender's cut(s) from each iteration and append all cuts to **RMP** directly. The way of adding cuts to **RMP** iteratively will quickly expand the size of the constraint set; however, it can't guarantee those newly generated cuts having stronger restriction on the solution space. In contrast, it is possible to bring in the considerable rework and slow down the convergence speed of the algorithm.

For the solution process, we can implement a tailored Bender's Decomposition in CPLEX by calling CALLBACK function. The solution flowchart (Fig. 3.4) explicitly addresses how to apply a modified Bender's Decomposition to solve two-stage stochastic UC models with the help of CALLBACK function. Different from the

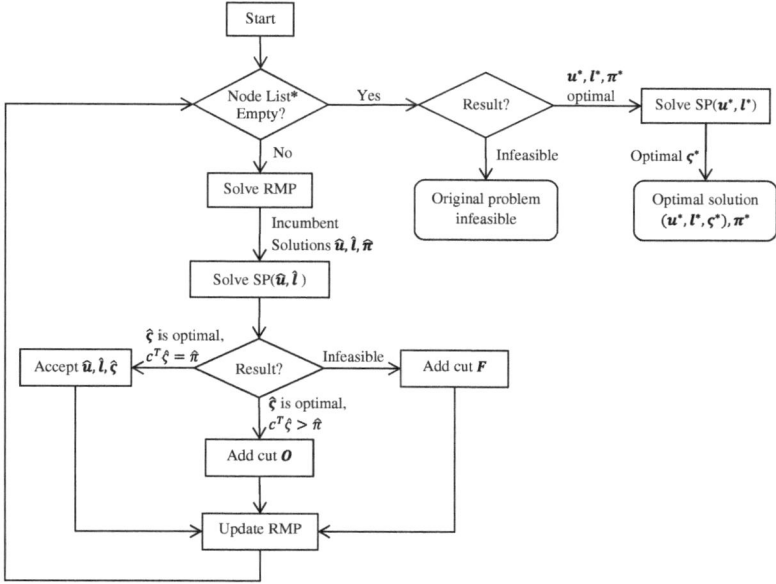

Fig. 3.4 The solution flowchart of Benders' Decomposition with CALLBACK function [HZW14]

traditional process, the Benders' decomposition with CALLBACK function has a major advantage that all cuts are carried in a pool, where only violated cuts will be appended to **RMP** in the following solving process. In doing so, the whole process is able to use a limited number of stronger or equivalent Benders's cuts and maintaining a small size of **RMP**.

Another significant advantage of using CALLBACK function is that **RMP** is solved only once. The whole process of solving **RMP** can be described by a Branch-and-Bound-and-Cut algorithm. One or more Benders' cuts are produced at a branching node, which is considered as the upper echelon of a hierarchical solution process. In fact, only the most violated Benders' cut(s) will be used finally at each node. The lower bound will be updated along with the **RMP** solving procedure in the branch-and-bound tree in which we can obtain the upper bound at a branching node after solving all SP^ξ, $\xi \forall \, \Xi$, in the lower echelon of hierarchical solution process. Until the **RMP** solving procedure is finished, if the difference between lower bound and upper bound is less than an expected error, we can say the algorithm get converged. Because the LB can be improved by Benders' cuts, it can avoid solving **RMP** iteratively without LB improvement in the traditional method, and also the overall computation time can be improved.

3.6 SUC with Real-Time Rescheduling

In real-time, the economic dispatch process aims to balance energy by dispatching imbalance energy, or the energy different from the schedule, and the energy requested from ancillary service. To avoid the imbalances, ISOs/RTOs initially perform reserve dispatches or rescheduling on online generating units. If additional units need to be synchronized, the rescheduling process ensures that the current committed units can't be desynchronized. On the real-time unit commitment, quick-start generators are designated in 15-min intervals and considered rescheduling in 15 min [Cal14]. These quick-start generators are usually gas turbine powered, but it can be expensive to dispatch these off-line or online (if there is much excessive electricity) generators. These generators are also regularly referred to as peakers. If the reserve and rescheduling are not able to provide enough generation resources for the demand, ISOs/RTOs consider resort to load shedding to secure the system's stability [OVK10]. In this section, we would like to study how to reschedule these quick-start units while knowing the day-ahead unit commitment schedule of all generation units.

3.6.1 Problem Formulation

In day-ahead unit commitment, both quick-start and non-quick-start generators are committed. In real time, only the quick-start generators can change from their day-ahead schedules. G_q is used to denote the set of quick-start generators. Let the rescheduled new status of the quick-start generators be denoted by β_{jt}^{ξ}. Note that this has a superscript ξ denoting the corresponding scenario. Because these units are agile, it does not need to have the minimum up and down constraints. However, it is important to model the changes made in the rescheduling so that we can incorporate the costs due to changes of commitments. Lets use binary variables v_{jt}^{ξ} and w_{jt}^{ξ} to denote the renewed start-up and shut-down actions respectively in real time for these quick-start generators. The constraints to define the actions are similar to day-ahead models, as follows,

$$v_{jt}^{\xi} \geq \beta_{jt}^{\xi} - \beta_{j(t-1)}^{\xi}, \quad \forall j \in G_q, t \in T, \xi \in \Xi$$
$$w_{jt}^{\xi} \geq \beta_{j(t-1)}^{\xi} - \beta_{jt}^{\xi}, \quad \forall j \in G_q, t \in T, \xi \in \Xi$$

These variables can help find the changes on startup and shutdown actions, which are respectively y_{jt}^{ξ} and z_{jt}^{ξ}. For example, in scenario ξ, $y_{jt}^{\xi} = 1$ means that generator j has a start-up action in real time which is not scheduled in the the day-ahead decisions; $y_{jt}^{\xi} = 0$ means that real-time action is as same as day-ahead schedule; $y_{jt}^{\xi} = -1$ tells that there is no start-up action the real time although the day-ahead schedule start this unit at time t. These relationships is modeled by the following constraints,

$$y_{jt}^{\xi} \geq v_{jt}^{\xi} - v_{jt}, \quad \forall j \in G_q, t \in T, \xi \in \Xi$$

$$z_{jt}^{\xi} \geq w_{jt}^{\xi} - w_{jt}, \quad \forall j \in G_q, t \in T, \xi \in \Xi$$

where z_{jt} is denoting the changes on shut-down actions of the quick-start generation units. The corresponding unit costs are CU_{jt} and CD_{jt} respectively for changes in start-up and shut-down actions. Other requirements and constraints are similar to the two-stage stochastic programming models for unit comment problems. For convenience, the sets (at time t) of minimum up times $(t, \ldots, \min\{t + L_i - 1, |T|\})$ and minimum down times $(t, \ldots, \min\{t + l_i - 1, |T|\})$ are denoted by $\mathcal{M}_{UP}(t)$ and $\mathcal{M}_{DWN}(t)$ respectively. In addition, let $\mathbf{u}, \mathbf{v}, \mathbf{w}$ be the vectors of day-ahead commitment status, startup actions, and shutdown actions respectively. Hence the two-stage stochastic unit comment with real-time unit rescheduling can be formulated as follows,

$$[\text{SUCRR}] : \min \sum_{t \in T} \sum_{i \in G} (SU_{it}v_{it} + SD_{it}w_{it}) + Q(\mathbf{u}, \mathbf{v}, \mathbf{w}) \tag{3.43}$$

$$\text{s.t. } u_{it} - u_{i(t-1)} \leq u_{i\tau}, \quad \forall i \in G, \tau \in \mathcal{M}_{UP}(t), t \in T, \tag{3.44}$$

$$u_{i(t-1)} - u_{it} \leq 1 - u_{i\tau}, \quad \forall i \in G, \tau \in \mathcal{M}_{DWN}(t), t \in T, \tag{3.45}$$

$$v_{it} \geq u_{it} - u_{i(t-1)}, \quad \forall i \in \{N_c \cup N_g\}, t \in T, \tag{3.46}$$

$$w_{it} \geq -u_{it} + u_{i(t-1)}, \quad \forall i \in \{N_c \cup N_g\}, t \in T, \tag{3.47}$$

$$u_{it}, v_{it}, w_{it} \in \{0, 1\}, \quad \forall i \in \{N_c \cup N_g\}, t \in T, \tag{3.48}$$

where $Q(\mathbf{u}, \mathbf{v}, \mathbf{w})$ is the value function of the day-ahead schedule. The first stage problem is as same as other standard two-stage stochastic unit commitment problems. The value function is defined as follows,

$$Q(\mathbf{u}, \mathbf{v}, \mathbf{w}) = \min \sum_{\xi \in \Xi} Prob^{\xi} \sum_{t \in T} \left[\sum_{i \in G} F_i \left(p_{it}^{\xi} \right) + \sum_{j \in G_q} \left(CU_{jt}y_{jt}^{\xi} + CD_{jt}z_{jt}^{\xi} \right) \right]$$
$$\tag{3.49}$$

$$\text{s.t. } \sum_{i \in G} p_{it}^{\xi} \geq D_t^{\xi}, \quad \forall t \in T, \xi \in \Xi, \tag{3.50}$$

$$\sum_{i \in G} s_{it}^{\xi} \geq RS_t^{\xi}, \quad \forall t \in T, \xi \in \Xi, \tag{3.51}$$

$$\sum_{j \in G_q} q_{jt}^{\xi} \geq RO_t^{\xi}, \quad \forall t \in T, \xi \in \Xi, \tag{3.52}$$

$$y_{jt}^{\xi} \geq v_{jt}^{\xi} - v_{jt}, \quad \forall j \in G_q, t \in T, \xi \in \Xi, \tag{3.53}$$

$$z_{jt}^{\xi} \geq w_{jt}^{\xi} - w_{jt}, \quad \forall j \in G_q, t \in T, \xi \in \Xi, \tag{3.54}$$

$$v_{jt}^{\xi} \geq \beta_{jt}^{\xi} - \beta_{j(t-1)}^{\xi}, \quad \forall j \in G_q, t \in T, \xi \in \Xi, \tag{3.55}$$

$$w_{jt}^{\xi} \geq -\beta_{jt}^{\xi} + \beta_{j(t-1)}^{\xi}, \quad \forall j \in G_q, t \in T, \xi \in \Xi, \tag{3.56}$$

$$P_i^{\max} u_{it} \geq p_{it}^{\xi} \geq P_i^{\min} u_{it}, \quad \forall i \in G \setminus G_q, t \in T, \xi \in \Xi, \tag{3.57}$$

$$p_{it}^{\xi} + s_{it}^{\xi} \leq P_i^{\max} u_{it}, \quad \forall i \in G \setminus G_q, t \in T, \xi \in \Xi, \tag{3.58}$$

$$P_j^{\max} \beta_{jt}^{\xi} \geq p_{jt}^{\xi} \geq P_j^{\min} \beta_{jt}^{\xi}, \quad \forall j \in N_g, t \in T, \xi \in \Xi, \tag{3.59}$$

$$p_{jt}^{\xi} + s_{jt}^{\xi} \leq P_j^{\max} \beta_{jt}^{\xi}, \quad \forall j \in G_q, t \in T, \xi \in \Xi, \tag{3.60}$$

$$- RD_i \leq p_{it}^{\xi} - p_{i(t-1)}^{\xi} \leq RU_i, \quad \forall i \in G, t \in T, \xi \in \Xi, \tag{3.61}$$

$$s_{it}^{\xi} \leq S_i^{\max}, \quad \forall i \in G, t \in T, \xi \in \Xi, \tag{3.62}$$

$$q_{jt}^{\xi} \leq (1 - \beta_{jt}^{\xi}) P_j^{\max}, \quad \forall j \in G_q, t \in T, \xi \in \Xi, \tag{3.63}$$

$$\beta_{jt}^{\xi}, v_{jt}^{\xi}, w_{jt}^{\xi} \in \{0, 1\}, \quad \forall j \in G_q, t \in T, \xi \in \Xi, \tag{3.64}$$

$$y_{jt}^{\xi}, z_{jt}^{\xi} \in \{-1, 0, 1\}, \quad \forall j \in G_q, t \in T, \xi \in \Xi, \tag{3.65}$$

$$\mathbf{p}^{\xi}, \mathbf{s}^{\xi}, \mathbf{q}^{\xi} \geq 0, \quad \forall \xi \in \Xi, \tag{3.66}$$

where $\mathbf{p}^{\xi}$, $\mathbf{s}^{\xi}$, $\mathbf{q}^{\xi}$ are the decision vectors for power dispatch, spinning reserve, and operating reserve respectively under scenarios ξ. The constraints in the value function are similar to the standard two-stage stochastic unit commitment except the constraints that model the changes of quick-start generator's commitment schedules, i.e., (3.53)–(3.56). These constraints introduce binary variables, which make the previously discussed algorithms (e.g., Benders decomposition, L-shaped methods) not applicable. Hence we need to resorts other techniques to decompose this large-scale optimization problems.

3.6.2 A Decomposition Algorithm

The optimization model introduced above was initially discussed by Zheng et al. [ZWPG13], where they also proposed an efficient decomposition algorithm to solve large-scale cases. In the following, the decomposition algorithm will be summarized and discussed. The central idea for decomposition of this stochastic mixed integer linear program is to approximate the value function $Q(\mathbf{u}, \mathbf{v}, \mathbf{w})$. As $Q(\cdot)$ is both noncontinuous and nonconvex, the algorithm use both combinatorial and dual cutting planes to ensure both the convergence and the convergent rate respectively.

To ensure convergence, Integer L-shaped cuts proposed in [LL93] is introduced to this particular problem as follows,

$$\pi \geq (Q(\hat{\mathbf{u}}, \hat{\mathbf{v}}, \hat{\mathbf{w}}) - L) \left(\sum_{(j,t) \in \mathbb{T}(\hat{u})} u_{jt} - \sum_{(j,t) \in \mathbb{F}(\hat{u})} u_{jt} - |\mathbb{T}(\hat{u})| + 1 \right) + L$$

where $Q(\hat{\mathbf{u}}, \hat{\mathbf{v}}, \hat{\mathbf{w}})$ is the value of function $Q(\cdot)$ at a given day-ahead schedule $\hat{\mathbf{u}}, \hat{\mathbf{v}}, \hat{\mathbf{w}}$, and L is a lower bound for $Q(\mathbf{u}, \mathbf{v}, \mathbf{w})$ given any day-ahead schedule, and $\mathbb{T}(\hat{\mathbf{u}})$ and $\mathbb{T}(\hat{\mathbf{u}})$ are the sets of the indices whose corresponding $\hat{\mathbf{u}}$ are "1" and "0" respectively. The reason that only $\mathbf{u}$ is considered in the cutting plane is that $\mathbf{v}$ and $\mathbf{w}$ are uniquely dependent on $\mathbf{u}$ in the optimal solution. Hence in the combinatorial cutting plane, there is only the day-ahead commitment decision variables but not start-up and shut-down action variables. In addition, if the current solution $(\hat{\mathbf{u}}, \hat{\mathbf{v}}, \hat{\mathbf{w}})$ is causing infeasibility of the second stage (or $Q(\hat{\mathbf{u}}, \hat{\mathbf{v}}, \hat{\mathbf{w}}) = \infty$), a feasibility combinatorial cuts should be added as follows,

$$\sum_{(j,t)\in\mathbb{T}(\hat{\mathbf{u}})} u_{jt} - \sum_{(j,t)\in\mathbb{F}(\hat{\mathbf{u}})} u_{jt} \leq |\mathbb{T}(\hat{\mathbf{u}})| - 1$$

Simply applying these Integer-L-Shaped cuts will eventually let the algorithm converge to the real optimal solution. However, it will be very slow as the combinatorial cutting plane will only remove one possible day-ahead schedule each time. Hence, as mentioned in [ZWPG13], Benders cuts are also added based on a nested approach. In this nested or embedded approach, the value function optimization problem is further decoupled to two parts in the following manner.

$$Q(\mathbf{u}, \mathbf{v}, \mathbf{w}) = \min \sum_{\xi\in\Xi} Prob^{\xi}\left[\sum_{i\in G}\mathsf{R}^{\xi}(\mathbf{u}, \beta^{\xi}) + \sum_{t\in T}\sum_{j\in G_q}\left(CU_{jt}y_{jt}^{\xi} + CD_{jt}z_{jt}^{\xi}\right)\right]$$
(3.67a)

$$\text{s.t.} \, y_{jt}^{\xi} \geq v_{jt}^{\xi} - v_{jt}, \quad \forall j \in G_q, t \in T, \xi \in \Xi, \tag{3.67b}$$

$$z_{jt}^{\xi} \geq w_{jt}^{\xi} - w_{jt}, \quad \forall j \in G_q, t \in T, \xi \in \Xi, \tag{3.67c}$$

$$v_{jt}^{\xi} \geq \beta_{jt}^{\xi} - \beta_{j(t-1)}^{\xi}, \quad \forall j \in G_q, t \in T, \xi \in \Xi, \tag{3.67d}$$

$$w_{jt}^{\xi} \geq -\beta_{jt}^{\xi} + \beta_{j(t-1)}^{\xi}, \quad \forall j \in G_q, t \in T, \xi \in \Xi, \tag{3.67e}$$

$$\beta_{jt}^{\xi}, v_{jt}^{\xi}, w_{jt}^{\xi} \in \{0, 1\}, \quad \forall j \in G_q, t \in T, \xi \in \Xi, \tag{3.67f}$$

$$y_{jt}^{\xi}, z_{jt}^{\xi} \in \{-1, 0, 1\}, \quad \forall j \in G_q, t \in T, \xi \in \Xi, \tag{3.67g}$$

where the value function $\mathsf{R}^{\xi}(\beta^{\xi})$ is defined as follows,

$$\mathsf{R}^{\xi}(\mathbf{u}, \beta^{\xi}) = \min \sum_{t\in T} F_i\left(p_{it}^{\xi}\right) \tag{3.68a}$$

$$\text{s.t.} \sum_{i\in G} p_{it}^{\xi} \geq D_t^{\xi}, \quad \forall t \in T, \xi \in \Xi, \tag{3.68b}$$

$$\sum_{i\in G} s_{it}^{\xi} \geq RS_t^{\xi}, \quad \forall t \in T, \xi \in \Xi, \tag{3.68c}$$

$$\sum_{j\in G_q} q_{jt}^{\xi} \geq RO_t^{\xi}, \quad \forall t \in T, \xi \in \Xi, \tag{3.68d}$$

$$P_i^{\max} u_{it} \geq p_{it}^{\xi} \geq P_i^{\min} u_{it}, \quad \forall i \in G \setminus G_q, t \in T, \xi \in \Xi,$$
$$\text{(3.68e)}$$

$$p_{it}^{\xi} + s_{it}^{\xi} \leq P_i^{\max} u_{it}, \quad \forall i \in G \setminus G_q, t \in T, \xi \in \Xi, \quad \text{(3.68f)}$$

$$P_j^{\max} \beta_{jt}^{\xi} \geq p_{jt}^{\xi} \geq P_j^{\min} \beta_{jt}^{\xi}, \quad \forall j \in N_g, t \in T, \xi \in \Xi, \text{(3.68g)}$$

$$p_{jt}^{\xi} + s_{jt}^{\xi} \leq P_j^{\max} \beta_{jt}^{\xi}, \quad \forall j \in G_q, t \in T, \xi \in \Xi, \quad \text{(3.68h)}$$

$$-RD_i \leq p_{it}^{\xi} - p_{i(t-1)}^{\xi} \leq RU_i, \quad \forall i \in G, t \in T, \xi \in \Xi,$$
$$\text{(3.68i)}$$

$$s_{it}^{\xi} \leq S_i^{\max}, \quad \forall i \in G, t \in T, \xi \in \Xi, \quad \text{(3.68j)}$$

$$q_{jt}^{\xi} \leq (1 - \beta_{jt}^{\xi}) P_j^{\max}, \quad \forall j \in G_q, t \in T, \xi \in \Xi, \quad \text{(3.68k)}$$

$$\mathbf{p}^{\xi}, \mathbf{s}^{\xi}, \mathbf{q}^{\xi} \geq 0, \quad \forall \xi \in \Xi, \quad \text{(3.68l)}$$

The optimal dual solutions of (3.68) will help construct a cutting plane for (3.67). Then the integrality constraints of (3.67) is relaxed to be able to obtain an approximated dual solution to (3.67), which is then used to generate a Benders cutting plane for the first-stage problem. The algorithm proposed in [ZWPG13] generates both these Benders' cutting planes and Integer-L-Shaped cuts at each iteration. The whole iteration process is as same as discussed previously in this book.

3.7 SUC with Contingency Management

The contingency management aims to cover transmission facilities outages and generation facilities outages. Apart from unpredictable and uncontrollable contingency, e.g. extreme weather-related blackouts, common contingencies are simulated forced outages of generators, network elements and their associated disconnect devices. The ISOs/RTOs coordinate and approve generator outage schedules; meanwhile, they process the requests for transmission outages that potentially affect the reliability of power systems [NYI13]. In addition, contingency management is comprised of determining the schedule of "opening"/"closing" one or more elements as well as the arrangement of the sequence of predefined contingency removal events.

The most contingencies probably have one or more element outage(s). As we state the operational solutions in Sect. 2.5, most unit commitment scheduling go through the network security check in order that no security violation is expected to occur when any forced outage contingency happens. When facing certain contingency conditions, the ISOs can utilize the reserves or adjust the dispatch for a short time interval, e.g. 10-min interval.

This section mainly introduces the modelling of contingency in the respective of generator outages and transmission outages under uncertainty.

3.7.1 Generating Unit Outage

A forced generation outage is viewed as a serious disturbance affecting the normal operation of system and its reliability. However, through approved generator outage schedules, it can avoid that the power system goes from normal operation status to emergency status. We thus need to consider the possibility of each generator outage happening in the system and incorporate the traditional UC model with explicit generator outage constraints in a stochastic environment. Following the stream of stochastic unit commitment, we introduce the two-stage stochastic unit commitment model with explicit reliability requirements.

The objective of SUC remains to minimize the operation costs and unserved energy penalty in a day-ahead market. However, in the first stage, a day-ahead reliability assessment commitment (RAC) is performed regarding unit commitment schedule and reserve commitment schedule; the second stage is to optimize next-day energy dispatch, reserve dispatch and power transmission based on all independent scenarios. The objective function (3.69) is similar to the previous SUC objective functions, but including the reserve costs of regulation services and related regulation dispatch costs.

$$
\begin{aligned}
\min \quad & \sum_{g \in G} \sum_{t \in T} (SU_{gt} v_{gt} + SD_{gt} w_{gt} + C_{gt}^{ru} rc_{gt}^{u} + C_{gt}^{rd} rc_{gt}^{d}) \\
& + \sum_{\xi \in \Xi} prob^{\xi} \sum_{t \in T} \sum_{g \in G} \left[F_g(p_{gt}^{\xi}) + F_r(r_{gt}^{u\xi}) + F_r(r_{gt}^{d\xi}) \right] \\
& + VOLL \sum_{\xi \in \Xi} prob^{\xi} \sum_{t \in T} \sum_{i \in N} \Delta_{it}^{\xi}
\end{aligned}
\tag{3.69}
$$

Note that the fuel cost is the quadratic function of the dispatch level, p, i.e., for generator g, $F_g(p) = a + bp + cp^2$, where a, b and c are usually positive coefficients. Similarly, the function $F_r(r_{gt}^{\xi})$ represent the costs of actually dispatching reserve resources in real time. The fuel cost of real-time regulation up(down) corresponds to the the quadratic function of the regulation level, $r^u(r^d)$, but with higher values of coefficients. Here we assume the regulation down service $r_{gt}^{d\xi}$ will incur cost. Due to the nonlinear objective function, a piecewise linear approximation is used to obtain very close solutions again.

In the first stage, unit commitment is scheduled according to the operation requirements for generating units such as minimum ON time, minimum OFF time, startup action and shutdown action. The regulation up and down reserves also are included to satisfy the forecasted reserve level in each period.

$$
u_{gt} - u_{g(t-1)} \leq u_{g\tau} \qquad \forall g \in G, \ t \in T,
$$
$$
\tau = t, \ldots, min\{t + L_g - 1\} \tag{3.70}
$$
$$
u_{g(t-1)} - u_{gt} \leq 1 - u_{g\tau} \qquad \forall g \in G, \ t \in T,
$$
$$
\tau = t, \ldots, min\{t + l_g - 1\} \tag{3.71}
$$

$$v_{gt} \geq u_{gt} - u_{g(t-1)} \qquad\qquad \forall g \in G, \, t \in T \qquad\qquad (3.72)$$

$$w_{gt} \geq -u_{gt} + u_{g(t-1)} \qquad\qquad \forall g \in G, \, t \in T \qquad\qquad (3.73)$$

$$\sum_{g \in G_i} rc_{gt}^{(u,d)} \geq R_{it}^{(u,d)} \qquad\qquad \forall i \in N, \, t \in T, \qquad\qquad (3.74)$$

$$u_{gt}, \, v_{gt}, \, w_{gt} \in \{0, 1\}, \qquad\qquad \forall g \in G, \, t \in T \qquad\qquad (3.75)$$

$$rc_{gt}^{u}, rc_{gt}^{d} \geq 0, \qquad\qquad \forall g \in G, \, t \in T \qquad\qquad (3.76)$$

The second stage constraints contain the economic dispatch including generation limits (3.77) and ramping limits (3.78)–(3.79), regulation up/down limits (3.80)–(3.81), and power transmission (3.82)–(3.83). Since the regulation up/down takes up a part of generation capacities when the units are ON, the ramping up/down is considered to cover both generation and regulation at the same moment. Any of generation changes or regulation changes can not exceed the ramp rate limit in successive periods. Meanwhile, constraints (3.80)–(3.81) ensure the real-time regulation up and down constrained by the regulation reserves requested from first stage. Additionally, constraints (3.82)–(3.83) show the traditional DC approximation of Kirchhoff's current law and Kirchhoff's voltage law applied into load balance, where the regulation up/down, renewable energy output and potential load-shedding loss are taken into account.

$$P_g^{min} u_{gt} \leq p_{gt}^{\xi} + r_{gt}^{u\xi} - r_{gt}^{d\xi} \leq P_g^{max} u_{gt}, \quad \forall \, g \in G, \, t \in T, \, \xi \in \varXi \quad (3.77)$$

$$(p_{gt}^{\xi} - p_{gt-1}^{\xi}) - (r_{gt}^{d\xi} - r_{gt-1}^{d\xi}) \geq -RD_g, \quad \forall \, g \in G, \, t \in T, \, \xi \in \varXi$$
$$(3.78)$$

$$(p_{gt}^{\xi} - p_{gt-1}^{\xi}) + (r_{gt}^{u\xi} - r_{gt-1}^{u\xi}) \leq RU_g, \quad \forall \, g \in G, \, t \in T, \, \xi \in \varXi \quad (3.79)$$

$$0 \leq r_{gt}^{u\xi} \leq rc_{gt}^{u}, \quad \forall \, g \in G, t \in T, \, \xi \in \varXi \qquad\qquad (3.80)$$

$$0 \leq r_{gt}^{d\xi} \leq rc_{gt}^{d}, \quad \forall \, g \in G, t \in T, \, \xi \in \varXi \qquad\qquad (3.81)$$

$$\sum_{(i,j) \in A_i^+} f_{ijt}^{\xi} - \sum_{(j,i) \in A_i^-} f_{jit}^{\xi} - \sum_{g \in G_i}(p_{gt}^{\xi} + r_{gt}^{u\xi} - r_{gt}^{d\xi}) - \Delta_{it}^{\xi}$$

$$= W_{it}^{\xi} - D_{it}^{\xi}, \quad \forall i \in N, \, t \in T, \, \xi \in \varXi \qquad\qquad (3.82)$$

$$(f_{ijt}^{\xi} - f_{jit}^{\xi}) - M_{ijt}^{\xi}(\beta_{it}^{\xi} - \beta_{jt}^{\xi}) = 0, \quad \forall \, (i, j) \in A, \, t \in T, \, \xi \in \varXi \quad (3.83)$$

$$-F_{ij}^{max} \leq f_{ijt}^{\xi} \leq F_{ij}^{max}, \quad \forall \, (i, j) \in A, \, i \in N, \, t \in T, \, \xi \in \varXi. \quad (3.84)$$

$$p_{gt}^{\xi} \geq 0, \quad \forall \, g \in G, \, t \in T, \, \xi \in \varXi \qquad\qquad (3.85)$$

$$\Delta_{it}^{\xi} \geq 0, \quad \forall \, i \in N, \, t \in T, \, \xi \in \varXi \qquad\qquad (3.86)$$

To model an outage state, we modified the generation limit constraint (3.77) to the generation availability constraint (3.87) with the consideration of generator outage situation. We denote the generator availability index α_{gt}^{ξ} for each scenario, when $\alpha_{gt}^{\xi} = 0$ indicates the generator outage occurs in scenario ξ at hour t; otherwise,

$\alpha_{gt}^{\xi} = 1$ and the constraint (3.87) is the same with (3.77).

$$\alpha_{gt}^{\xi} P_g^{min} u_{gt} \leq p_{gt}^{\xi} + r_{gt}^{u\xi} - r_{gt}^{d\xi} \leq \alpha_{gt}^{\xi} P_g^{max} u_{gt},$$
$$\forall g \in G, \ t \in T, \ \xi \in \Xi \qquad (3.87)$$

3.7.2 Transmission Outage

The ISOs can access transmission information from TRANSCOs and perform congestion management as well as contingency management. Another purpose of congestion and contingency management are to minimize transmission flow violations and risks of supply-demand imbalance. Compared to transmission congestion, the outcomes due to contingency are usually more serious. Therefore, ISOs expect to offer sufficient ancillary services to maintain the reliability of power system operations, taking into account of $N - 1$ contingency conditions.

In day-ahead or hour-ahead power markets, ancillary services such as regulation, spinning reserve, non-spinning reserve, or operating reserve are procured to prevent one of outcomes from transmission line outages. This way requires a large amount of reserve and holds a lot of generating resources.

Stochastic programming also provides a great help in effectively modelling the transmission contingency under uncertainties. For the simplicity of modeling a contingency problem, the common transmission outages including line and transformer are simulated independently; meanwhile, the physical locations of contingency are explicitly taken into account. Random outage occurs at any transmission line (i, j) or transformer k between any two time period(s) $[t, \ t + n]$. We denotes a binary parameter $N1_{ij} = 1$ for the contingency on transmission line, and a binary parameter $N1_k = 1$ for the contingency on transformer. Then the formulations for transmission line outages are presented on line and transformer separately, and include transmission flow capacities (3.88), transformer status (3.88) and phase angles (3.90)–(3.92).

$$-F_{ij}^{max} N1_{ijt}^{\xi} \leq f_{ijt}^{\xi} \leq F_{ij}^{max} N1_{ijt}^{\xi}, \quad \forall (i, j) \in A, \ t \in T, \ \xi \in \Xi \qquad (3.88)$$
$$F_{ij}^{min} N1_{kt}^{\xi} \leq F_{kt}^{\xi} \leq F_{ij}^{max} N1_{kt}^{\xi}, \quad \forall (i, j) \in A, \ t \in T, \ \xi \in \Xi \qquad (3.89)$$
$$(f_{ijt}^{\xi} - f_{jit}^{\xi}) - B_{ijt}(\beta_{it}^{\xi} - \beta_{jt}^{\xi}) + (1 - N1_{ij})M_{ij} \geq 0, \qquad (3.90)$$
$$\forall (i, j) \in A, \ t \in T, \ \xi \in \Xi \qquad (3.91)$$
$$(f_{ijt}^{\xi} - f_{jit}^{\xi}) - B_{ijt}(\beta_{it}^{\xi} - \beta_{jt}^{\xi}) - (1 - N1_{ij})M_{ij} \leq 0, \qquad (3.92)$$
$$\forall (i, j) \in A, \ t \in T, \ \xi \in \Xi \qquad (3.93)$$

where F_{kt}^{ξ} is real power flow for transformer k at time t. F_{ij}^{min} and F_{ij}^{max} are the maximum and minimum ratings of transformer k, respectively.

Note that these formulations are correct in theory, but it may easily cause a whole system violation if a single contingency occurs. This situation is attributed to the high penalty costs or the infeasibility on optimization problems. This also implies the current system fragile with less agility. The main reasons are that a line outage causes the rest of line capacities fail to meet the required transmission amount or breaks the only connection between two buses.

To ensure transmission flow security in $N - 1$ contingency cases, recent studies focused on the optimal transmission switching (TS) [HOFO09], the co-optimization of unit commitment and transmission switching [HFO+10, KS10], and the optimal long-term generation and transmission outage scheduling with short-term UC [WSF10].

The transmission switching embedded in UC problem is formulated for the $N - 1$ compliant systems, in which a power system is able to survive due to the loss of any single network component (generator or transmission element). Through the co-optimization of TS and UC, one can investigate how TS affects UC when a contingency occurs and vice versa. We thus provide a two-stage SUC model for this co-optimization of TS and UC.

$$
\min \sum_{g \in G} \sum_{t \in T} (SU_{gt} v_{gt} + SD_{gt} w_{gt}) + \sum_{\xi \in \Xi} Prob^{\xi} \sum_{t \in T} \sum_{g \in G} [(b_{gt} p_{gt}^{\xi} + a_{gt} u_{gt})]
$$
$$
+ \sum_{\xi \in \Xi} Prob^{\xi} VOLL \sum_{t \in T} \sum_{i \in N} \Delta_{it}^{\xi}
$$

s.t. UC constraints

$$
P_g^{min} u_{gt} N1_g^{\xi} \leq p_{gt}^{\xi} \leq P_g^{max} u_{gt} N1_g^{\xi}, \quad \forall g \in G,\, t \in T,\, \xi \in \Xi \tag{3.94}
$$

$$
- RD_g \leq p_{gt}^{\xi} - p_{gt-1}^{\xi} \leq RU_g, \quad \forall g \in G,\, t \in T,\, \xi \in \Xi
$$

$$
\sum_{(i,j) \in A_i^+} f_{ijt}^{\xi} - \sum_{(j,i) \in A_i^-} f_{jit}^{\xi} - \sum_{g \in G_i} p_{gt}^{\xi} - \Delta_{it}^{\xi} = W_{it}^{\xi} - D_{it},\, \forall i \in N,\, t \in T,\, \xi \in \Xi
$$

$$
(f_{ijt}^{\xi} - f_{jit}^{\xi}) - B_{ijt}(\beta_{it}^{\xi} - \beta_{jt}^{\xi}) + (2 - z_{ijt} - N1_{ij}) M_{ij} \geq 0,
$$
$$
\forall (i, j) \in A,\, t \in T,\, \xi \in \Xi \tag{3.95}
$$

$$
(f_{ijt}^{\xi} - f_{jit}^{\xi}) - B_{ijt}(\beta_{it}^{\xi} - \beta_{jt}^{\xi}) - (2 - z_{ijt} - N1_{ij}) M_{ij} \leq 0,
$$
$$
\forall (i, j) \in A,\, t \in T,\, \xi \in \Xi \tag{3.96}
$$

$$
- F_{ij}^{max} N1_{ij}^{\xi} z_{ijt} \leq f_{ijt}^{\xi} \leq F_{ij}^{max} N1_{ij}^{\xi} z_{ijt}, \quad \forall (i, j) \in A,\, t \in T,\, \xi \in \Xi \tag{3.97}
$$

$$
\beta^{min} \leq \beta_{it}^{\xi} \leq \beta^{max}, \quad \forall i \in N,\, t \in T,\, \xi \in \Xi \tag{3.98}
$$

$$
\Delta_{it}^{\xi} \geq 0, \quad \forall i \in N,\, t \in T,\, \xi \in \Xi
$$

$$
f_{ijt}^{\xi}, \quad \forall (i, j) \in \mathscr{A},\, t \in T
$$

where constraints (3.95)–(3.97) specify the line switching decisions that affect the phase angles and the transmission flows in a network under the $N - 1$ contingency uncertainty.

When a transmission line is opened (not in service), constraints (3.95) and (3.96) are satisfied no matter what values are corresponding to phase angles. Under this situation, the transmission line also can be opened as a result of transmission switching, $z_{ijt} = 0$. The parameter M_{ij} is a "big M" value used to make the constraint non-binding, indicating that $M_{ij} \geq |B_{ijt}(\beta_{it} - \beta_{jt})|$. When either $z_{ijt} = 0$ or $N1_{ijt} = 0$, power flow f_{ijt} becomes zero, and then the function of M_{ij} ensures that constraints (3.95) and (3.96) hold regardless of the difference in phase angles.

3.8 Two-Stage SUC with Risk Constraints

We know that an attractive feature of stochastic unit commitment models that can take into account some certain uncertainties in power systems and their resulting risks. In addition to considering operational economics, the risk mentioned in this section refers to the likelihood of a load-shedding loss that happens at a bus given a time period.

The common method to handle load losses has been discussed in Sect. 2.3.6 through unserved energy constraint. This method is a "hard" way to simply restrict the hourly system-level loss to a predetermined loss allowance, neglecting which uncertainty factor induces the load loss and the probability of occurrence. Once we can identify significant factors bring uncertainties or disturbances to the power system, we can build a stochastic optimization model using either stochastic programming or robust programming and denote the load loss as a scenario-based variable to capture the effects of uncertainty sources.

Here we use a stochastic representation of renewable output, nodal demand, fuel price or all above. In order to describe an uncertainty more accurately, we usually produce a large number of simulated scenarios to support the SUC modeling, particularly including some extreme cases. This may lead to an optimal solution become overconservative, and also the recommended solution could come with high operational costs in that any operational solutions have to compensate a lot for handling extreme scenarios. Actually, to make an operational schedule, it's more reasonable to base on the majority of common scenarios with high probabilities and secure the system reliability with a target level of risk reduction. For a tradeoff between operational cost and system reliability, we introduce two popular risk constraints that have been applied in two-stage SUC models and address how they can be incorporated with UC models to conditionally control the risk of load loss.

3.8.1 Value at Risk

We first discuss how the loss of load to be linked with the risk in SUC problems. As a reliable system should be able to meet as much demand as it can in an interrupted condition, we use Loss of Load Expectation (LOLE), or Loss of Load Probability (LOLP), to indicate the amount of installed generation capacity required and thus

LOLE can stay below a specific level to meet a desired reliability target. More details about LOLE and LOLP can be accessed in [OMN04, WWG13a].

To come up with a robust scheduling, a power system should have enough generation capacities to satisfy any load with the aids of regulation service, shown as

$$\sum_{g \in G} p_{gt} - r_{gt}^d \leq D_{it} \leq \sum_{g \in G} p_{gt} + r_{gt}^u, \quad \forall i \in N.$$

Note that the forecasted demand can be replaced by a net load which is defined by $\Lambda_{it}^0 = D_{it}^0 - R_{it}^0$. If demand and renewable energy output are described by Normal distribution, the net load deviation is expressed by $\sigma_{it}^2 = (\sigma_{it}^D)^2 + (\sigma_{it}^R)^2$. Additionally, the generation capacities can be expended including operating reserve and non-generation resources.

In a real world, there is a certain possibility that both scheduled thermal generation and reserves fail to satisfy demands. For this case, ISOs would implement load shedding to satisfy unmet demands or renewable energy curtailment to avoid power overproduction, especially for wind generation. These two operations are not triggered frequently only when the power system encounters the following net load situations. The probability of either situation is usually in the two tails of Normal distribution.

- Case 1: High demands and low renewable energy outputs

This situation will make ISOs deploy the upward regulation and the spinning reserve to satisfy demands, and the unserved energy can be expressed by a reliability distribution function Υ_{it}^ξ, given as

$$\Upsilon_{it}^\xi = D_{it}^\xi - \sum_{g \in G_i} p_{gt}^\xi + (r_{gt}^u)^\xi, \quad \forall i \in N, \ t \in T, \ \xi \in \Xi.$$

When $\Upsilon_{it}^\xi \leq 0$, the system has no risk for any scenario. When $\Upsilon_{it}^\xi > 0$, the load shedding is executed at some buses. The corresponding possibility of occurrence is defined as

$$LOLP_{it} = \text{Prob}\left\{\Upsilon_{it}^\xi > 0\right\}, \quad \forall \ i \in N, \ t \in T,$$

and the expected unserved energy is defined as

$$EUE_{it} = \mathbb{E}\left[\Upsilon_{it} | \Upsilon_{it}^\xi > 0\right], \quad \forall i \in N, \ t \in T.$$

- Case 2: Low demands and high renewable energy outputs

Similar to case I, the renewable energy curtailment (i.e., wind curtailment) is used to handle the abundant renewable energy generation and reduce their impacts on real-time power balance and thermal energy generation. For this situation, after running the downward regulation, we define another reliability distribution function Ψ_{it}^ξ given as

$$\Psi_{it}^{\xi} = D_{it}^{\xi} - \sum_{g \in G_i} p_{gt}^{\xi} - (r_{gt}^d)^{\xi}, \quad \forall \ i \in N, \ t \in T, \ \xi \in \Xi.$$

When $\Psi_{it}^{\xi} \geq 0$, the system is considered to be non-risky for any scenario. When $\Psi_{it}^{\xi} < 0$, the renewable energy curtailment will be performed after the completion of regulation down. The probability and the expected renewable energy curtailment are respectively defined as follow

$$LORP_{it} = \mathrm{Prob}\left\{\Psi_{it}^{\xi} < 0\right\}, \quad \forall i \in N, \ t \in T,$$

and

$$ERC_{it} = -\mathbb{E}\left[\Psi_{it}|\Psi_{it}^{\xi} < 0\right], \quad \forall i \in N, \ t \in T.$$

The LOLP in nature is similar to the chance constraint bounding the risk of load loss. More specifically, the risk of load loss can be presented by a θ-level Value at Risk (VaR) of the loss of load, where θ is a confidence level.

On one side, to merge the LOLP into constraints, we have several ways to aggregate the loss of load (e.g., the total losses over a specific planning horizon vs. the loss at one time period), and these methods mainly rely on the degree of risk control in a power system [WGW12].

On the other side, we can employ the Value-at-Risk (VaR) measure to quantify the risk of load loss. Let $L(x, Y)$ denote the load loss function (assuming the aggregated system-level loss at a time period), where x are the aggregated decision vector and Y is the stochastic input vector (e.g., demand and renewable energy output). $\mathrm{VaR}_{\theta}[L(x, Y)]$ is an abstract form to express the θ-level VaR of the loss of load function $L(x, Y)$. It also can be considered as a loss at the θ-level percentile of the distribution of losses over a planning horizon, shown as follows.

$$\mathrm{VaR}_{\theta}[L(x, Y)] = \min_{l}\left\{l\middle|\ \mathrm{Prob}\Big(L(x, Y) \leq l\Big) \geq \theta\right\}.$$

In chance constraints, $\mathrm{VaR}_{\theta}[L(x, Y)]$ is bounded above by the maximum acceptable loss of load $\bar{l}$. If a system has a high level of reliability target, the value of loss allowance $\bar{l}$ will restricted to close to zero.

From the definition of VaR, $\mathrm{VaR}_{\theta}[L(x, Y)]$ is generally nonconvex with respect to $L(x, Y)$; in other words, the constraints $\mathrm{VaR}_{\theta}[L(x, Y)] \leq l$ and Prob $\{L(x, Y) \leq l\} \geq \theta$ could be nonconvex. As there are binary variables in the VaR constraints and a big M used for good/bad scenario selection, the non-convexity indeed brings many computational challenges, particularly when if a chance-constrained problem involves a large number of scenarios and multiple uncertainty factors. Some researchers have attempted to use tailored approximation algorithms (i.e. Sample Average Approximation) to solve such type of chance-constrained SUC problems [WGW12, WWG13a].

3.8.2 Conditional Value at Risk

Conditional Value at Risk (CVaR) is an extension of VaR measure and an alternative methodology to quantifies and manages the risks of load loss. It's also referred as Average Value at Risk (AVaR) or Expected Tail Loss (ETL) in Finance area. The CVaR works as a percentile measure of risk extensively used in energy related areas, such as natural gas system expansion planning [ZP10], power trading in day-ahead energy market [DFTC09], stochastic network optimization [ZSS15], home energy management system [WZLZ14].

The CVaR of $L(x, Y)$ with confidence level $\theta \in [0, 1]$ is the conditional expectation of the loss function given that the loss exceeds $\text{VaR}_\theta[L(x, Y)]$. Thus, an optimal solution of CVaR-based SUC model contains the information of VaR measure. By the definition in [SSU08], the CVaR constraints are expressed through the VaR definition as follows,

$$\min_l \left\{ l \mid \text{Prob}\Big(L(x, Y) \le l\Big) \ge \theta \right\} = \eta \tag{3.99a}$$

$$\mathbb{E}\Big[L(x, Y) \big| L(x, Y) \ge \eta\Big] \le \bar{\phi} \tag{3.99b}$$

where η is equivalent to $\text{VaR}_\theta[L(x, Y)]$, $\mathbb{E}[\cdot]$ refers to the load loss expectation when the loss exceeds VaR, and $\bar{\phi}$ is the maximum value of loss allowance for CVaR. Piratically, we can estimate a value for loss allowance by a given system-level loss limit,

$$\bar{\phi} = [Loss\ Limit/(Max\ Total\ Demand)] \times 100\%.$$

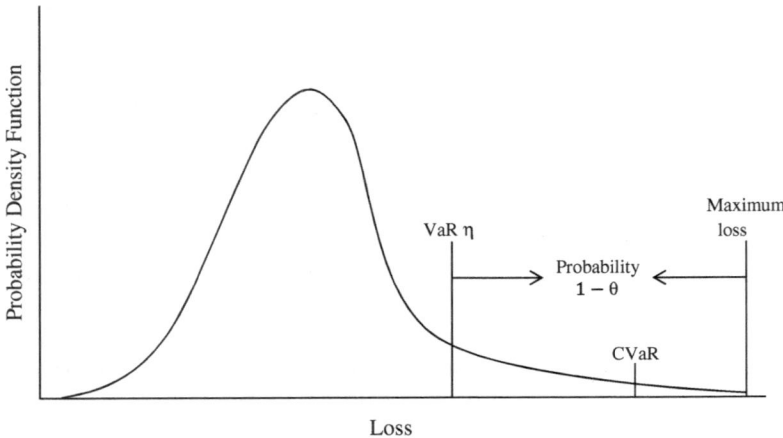

Fig. 3.5 VaR and CVaR on loss [SSU08]

Figure 3.5 shows the calculation of CVaR from the probability distribution of load-shedding loss. In addition to the VaR constraint (3.99a), Constraint (3.99b) states an additional requirement that the expectation of the losses exceeding VaR should be not greater than the loss allowance $\bar{\phi}$. Only the load loss falling between VaR and the Maximum loss will be captured in Constraint (3.99b). Regarding the relationship between VaR and CVaR on loss control, one may note that the VaR is always a loss less than or equal to the CVaR. Readers can obtain more details for these two measure techniques from [RU00, SSU08].

For employing CVaR constraints into a SUC problem, we need to determine if CVaR is used to link to the system-level load loss for a single time period or the load loss at a specific bus over all time periods. For instance, we model the system loss expectation beyond $\text{VaR}_\theta[L(x, Y)]$, and the loss in time period t is divided into two parts, η_t and ζ_t^ξ as shown in (3.100). Variable η_t represents the calculated $\text{VaR}_\theta[L(x, Y)]$ at time t, and ζ_t^ξ states the loss beyond the VaR in scenario ξ, but less than $\bar{\phi}$. They are defined as non-negative variables for indicating the losses in an independent scenario. The left hand side of constraint (3.101) corresponds to the expected system-level losses on (3.99b). In other words, these two terms are used for the calculation of the conditional loss expectation, which is $\text{CVaR}_\theta[L(x, Y)]$. Then this CVaR_θ is limited by a system loss allowance parameter $\bar{\phi}$ or hourly loss allowance $\bar{\phi}_t, \forall t \in T$. A group of CVaR-based risk constraints represent as follows,

$$\sum_{i \in N} l_{it}^\xi \leq \eta_t + \zeta_t^\xi, \qquad\qquad \forall t \in T, \xi \in \Xi \qquad (3.100)$$

$$\eta_t + (1 - \theta)^{-1} \sum_{\xi \in \Xi} \text{Prob}^\xi \zeta_t^\xi \leq \bar{\phi}_t, \qquad\qquad \forall t \in T \qquad (3.101)$$

$$\eta_t \geq 0, \ \zeta_t^\xi \geq 0, \qquad\qquad \forall t \in T, \xi \in \Xi \qquad (3.102)$$

It should be noted that CVaR constraints are shown above only involving continuous variables and linear constraints. Because of convexity, this makes the CVaR application very attractive in stochastic optimization models, especially when solving a model for a large-scale power system with a large number of scenarios.

3.8.3 Case 4: Two-Stage Stochastic Unit Commitment with CVaR Risk Constraints

This case study is to provide an example of a two-stage stochastic unit commitment model incorporating with conditional value-at-risk constraints (SUC-CVaR) for controlling the risk of load loss. The following optimization model and result analysis can help readers to understand how the reliability decision parameters in CVaR constraints can affect the operations. The two-stage SUC-CVaR model is given as

$$\min \sum_{g \in G} \sum_{t \in T} (SU_{gt} v_{gt} + SD_{gt} w_{gt}) + \sum_{\xi \in \Xi} Prob^{\xi} \sum_{t \in T} \sum_{g \in G} [(b_{gt} p_{gt}^{\xi} + a_{gt} u_{gt})]$$

s.t. $u_{gt} - u_{g(t-1)} \leq u_{g\tau}, \quad \forall g \in G, \ t \in T, \tau = t, \ldots, min\{t + L_g - 1, |T|\}$

$\quad u_{g(t-1)} - u_{gt} \leq 1 - u_{g\tau}, \quad \forall g \in G, \ t \in T, \tau = t, \ldots, min\{t + l_g - 1, |T|\}$

$\quad v_{gt} \geq u_{gt} - u_{g(t-1)}, \quad \forall g \in G, \ t \in T$

$\quad w_{gt} \geq -u_{gt} + u_{g(t-1)}, \quad \forall g \in G, \ t \in T$

$\quad u_{gt}, v_{gt}, w_{gt} \in \{0, 1\}, \quad \forall g \in G, \ t \in T$

$\quad P_g^{min} u_{gt} \leq p_{gt}^{\xi} \leq P_g^{max} u_{gt}, \quad \forall g \in G, \ t \in T, \ \xi \in \Xi$

$\quad -RD_g \leq p_{gt}^{\xi} - p_{gt-1}^{\xi} \leq RU_g, \quad \forall g \in G, \ t \in T, \ \xi \in \Xi$

$\quad s_{gt}^{\xi} \leq S_g^{max}, \quad \forall g \in G, t \in T, \ \xi \in \Xi$

$\quad \sum_{g \in G_i} s_{gt}^{\xi} \geq RS_{it}, \quad \forall i \in N, t \in T, \ \xi \in \Xi$

$\quad \sum_{(i,j) \in A_i^+} f_{ijt}^{\xi} - \sum_{(j,i) \in A_i^-} f_{jit}^{\xi} - \sum_{g \in G_i} p_{gt}^{\xi} - l_{it}^{\xi} = W_{it}^{\xi} - D_{it}^{\xi},$

$$\forall i \in N, \ t \in T, \ \xi \in \Xi$$

$\quad \sum_{i \in N} l_{it}^{\xi} \leq \eta_t + \zeta_t^{\xi}, \quad \forall t \in T, \ \xi \in \Xi$

$\quad \eta_t + (1 - \theta)^{-1} \sum_{\xi \in \Xi} Prob^{\xi} \zeta_t^{\xi} \leq \bar{\phi}, \quad \forall t \in T$

$\quad \eta_t \geq 0, \ \zeta_t^{\xi} \geq 0, \quad \forall t \in T, \ \xi \in \Xi$

$\quad p_{gt}^{\xi}, s_{gt}^{\xi} \geq 0, \quad \forall g \in G, \ t \in T, \ \xi \in \Xi$

$\quad \Delta_{it}^{\xi} \geq 0, \quad \forall i \in N, \ t \in T, \ \xi \in \Xi$

$\quad f_{ijt}^{\xi}, \quad \forall (i, j) \in \mathscr{A}, \ t \in T$

We solve the 7-bus system problem through the SUC-CVaR model without any assistance of non-generation resources, rescheduling or contingency management. Also, we discuss the effects of reliability parameters on the total operational costs, based on a 100-wind-scenario case.

For simplicity, we assume that the fuel cost function is linear regardless of the quadratic term in the objective function. Firstly, we run the SUC-CVaR model given with 85% confidence level and 5% of maximum hourly demands as load-shedding loss allowance. The optimal unit commitment schedule and its total cost are shown in Table 3.3. It can be seen that the SUC-CVaR solution keeps G1, G2 and G4 running a whole day, but G3 is turned off because of its high fuel cost as well as high startup/shutdown cost. Although G4 has the second most expensive fuel cost generation, the SUC-CVaR solution recommends G4 online running at a low production level to accommodate some demand changes. Because of its high ramping rate, it's

Table 3.3 Optimal unit commitment for a 7-bus system

Model type	Operational cost	Unit ID	Hour (1–24)
SUC-CVaR	$54917.9	G1	1 1 1 1 1 1 1 1 1 1 1 1 1 1 1 1 1 1 1 1 1 1 1 1
		G2	1 1 1 1 1 1 1 1 1 1 1 1 1 1 1 1 1 1 1 1 1 1 1 1
		G3	0 0 0 0 0 0 0 0 0 0 0 0 0 0 0 0 0 0 0 0 0 0 0 0
		G4	1 1 1 1 1 1 1 1 1 1 1 1 1 1 1 1 1 1 1 1 1 1 1 1

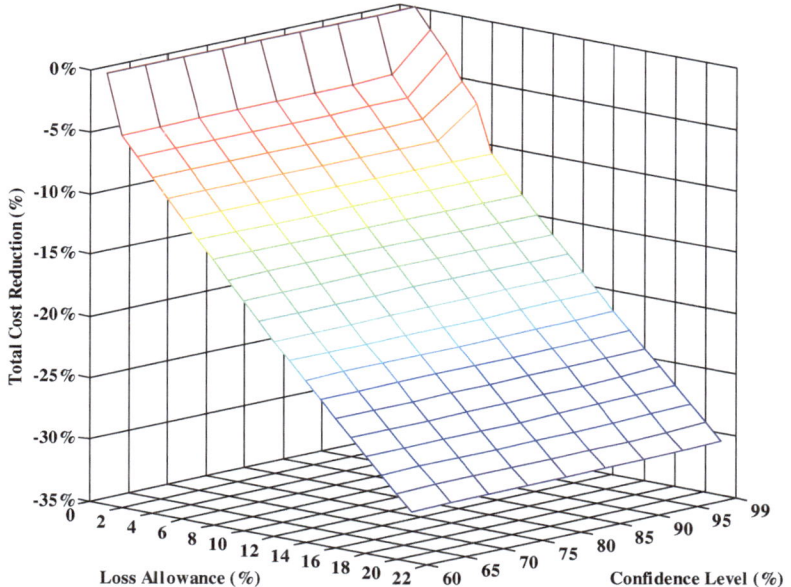

Fig. 3.6 Cost saving comparisons in three-dimension (A 7-bus system)

more appropriate to use G4 to deal with the volatility of wind energy generation in the way of increasing/decreasing generation outputs frequently.

Next, we performed a set of numerical tests using different reliability parameter settings in CVaR constraints. The sensitivity analysis is used to tell us how the operational cost saving can be attributed to the θ confidence level and the loss allowance $\bar{\phi}$. We can see that the cost saving is significant when raising the confidence level θ from 60 to 99% and the load-shedding loss allowance $\bar{\phi}$ from 1 to 20%.

Figure 3.6 shows the cost saving reduction with respect to different confidence levels and loss allowances, respectively. Two horizontal axes represents the decision input parameters, θ and $\bar{\phi}$, and the vertical axis stands for a percent of total operational cost reductions. One numerical point stands in an intersection on the plane. When no loss is allowed (i.e. $\bar{\phi} = 0$), the total operational cost is most expensive since the operation solution has to make sure all demands to be satisfied in any scenario. Here

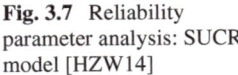

Fig. 3.7 Reliability
parameter analysis: SUCR
model [HZW14]

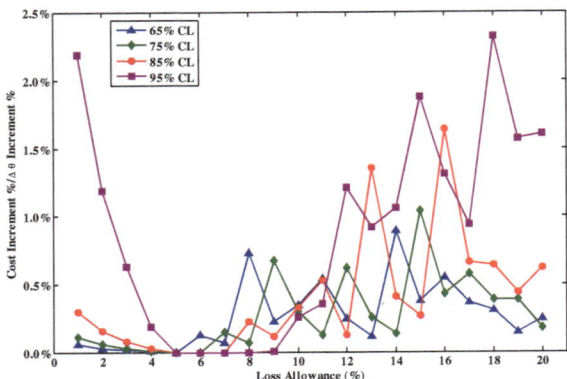

we take the point at the 0% of loss allowance, the 99% of confidence level and the
0% of cost saving reduction as a benchmark for the optimal cost comparisons.

As θ increases, the total operational cost appears constantly decreasing throughout
all investigated confidence levels. At the point when $\bar{\phi} = 20\%$ and $\theta = 60\%$, we
can get the largest cost reduction, but this solution may be not suitable for a real
power system as this loss allowance is too high and lowers the system reliability
and customer satisfaction. We also see a big drop over 5% reduction when loss
allowance increasing from 0 to 2%. As for the 99% of confidence level, the impacts
of loss allowance change on total cost saving become less until the loss allowance
raises from 8 to 10%.

Here we also provide an example to show the reliability parameter analysis which
is very helpful to identify a specific range highly affected by loss allowance. Figure 3.7
shows the relationship between the change rate of total cost increment (i.e., $\frac{\partial z^*}{\partial \theta}(\bar{\phi})$)
and the loss allowance. As for the representation of cost increment, we are interested
in the percentage of cost increment given with the percent difference of θ increment,
described in the y-axis. We can observe that each confidence level has its own volatil-
ities. If we increase a loss allowance within a range [8% 20%], it shows that the total
cost increment may vary a lot and typically become more volatile for all confidence
levels, compared to another lower level of loss allowance (<8%).

3.9 Summary

This chapter mainly introduces the two-stage stochastic programming method to
solve unit commitment problems under uncertainty, in terms of incorporating with
demand response, energy storage, real-time rescheduling, outage management, risk
control and aversion. The participation of non-generation resources can make renewable energy more assessable and mitigate the uncertainties from supply and demand
sides. The optimization of real-time rescheduling further improves the economic

dispatch process, and more importantly, ensure to adopt power supply changes in addition to from day-ahead schedule or from ancillary services. The contingency-constrained UC focuses on the optimal scheduling with the considerations of forced generator or transmission outage. As to securing the system reliability, the operation schedule would become more conservative and costly. When unexpected contingency happens or the total demand exceeds the generation or transmission capacities during a time period, load shedding is the last-resort measure to resolve energy imbalance. Hence the load-shedding risk control is also required in the way of applying LOLP or CVaR constraints into stochastic UC models.

References

[AJ09]   Abbey C, Joós G (2009) A stochastic optimization approach to rating of energy storage systems in wind-diesel isolated grids. IEEE Trans Power Syst 24(1):418–426
[Cal14]   California Independent System Operator (2014) California ISO - Market Processes. http://www.caiso.com/market/Pages/MarketProcesses.aspx
[CTT+09]   Chakraborty S, Senjyu T, Toyama H, Saber A, Funabashi T (2009) Determination methodology for optimising the energy storage size for power system. IET Gener Transm Distrib 3(11):987–999
[DFTC09]   Dicorato M, Forte G, Trovato M, Caruso E (2009) Risk-constrained profit maximization in day-ahead electricity market. IEEE Trans Power Syst 24(3):1107–1114
[DLOR12]   Dietrich K, Latorre J, Olmos L, Ramos A (2012) Demand response in an isolated system with high wind integration. IEEE Trans Power Syst 27(1):20–29
[HFO+10]   Hedman KW, Ferris MC, O'Neill RP, Fisher EB, Oren SS (2010) Co-optimization of generation unit commitment and transmission switching with N-1 reliability. IEEE Trans Power Syst 25(2):1052–1063
[HOFO09]   Hedman KW, O'Neill RP, Fisher EB, Oren SS (2009) Optimal transmission switching with contingency analysis. IEEE Trans Power Syst 24(3):1577–1586
[HZW14]   Huang Y, Zheng QP, Wang J (2014) Two-stage stochastic unit commitment model including non-generation resources with conditional value-at-risk constraints. Electr Power Syst Res 116:427–438
[JDBD11]   Jonghe CD, Delarue E, Belmans R, D'haeseleer W (2011) Determining optimal electricity technology mix with high level of wind power penetration. Appl Energy 88(6):2231–2238
[KM11]   Kowli AS, Meyn SP (2011) Supporting wind generation deployment with demand response. In: IEEE power and energy society general meeting, pp 1–8
[KS10]   Khodaei A, Shahidehpour M (2010) Transmission switching in security-constrained unit commitment. IEEE Trans Power Syst 25(4):1937–1945
[LL93]   Laporte G, Louveaux FV (1993) The integer L-shaped methods for stochastic integer programs with complete recourse. Oper Res Lett 13:133–142
[NYI13]   NYISO Energy Market Operations (2013) Outage Scheduling Manual. New York Independent System Operator, 4.1 edn
[OMN04]   Ozturk UA, Mazumdar M, Norman BA (2004) A solution to the stochastic unit commitment problem using chance constrained programming. IEEE Trans Power Syst 19(3):1589–1598
[OVK10]   Ortega-Vazquez MA, Kirschen DS (2010) Assessing the impact of wind power generation on operating costs. IEEE Trans Smart Grid 1(3):295–301
[RLL08]   Rong A, Lahdelma R, Luh PB (2008) Lagrangian relaxation based algorithm for trigeneration planning with storages. Eur J Oper Res 188(1):240–257

[RU00]   Rockafellar RT, Uryasev S (2000) Optimization of conditional value-at-risk. J Risk
              2:21–41
[Sio10]   Sioshansi R (2010) Evaluating the impacts of real-time pricing on the cost and value
              of wind generation. IEEE Trans Power Syst 25(2):741–748
[SSU08]   Sarykalin S, Serraino G, Uryasev S (2008) Value-at-risk vs. conditional value-at-risk
              in risk management and optimization. In: Tutorials in operations research. INFORMS,
              pp 270–294
[U.S13]   U.S. Department of Energy (2013) Grid energy storage. Technical report, U.S. Depart-
              ment of Energy
[WGW12]   Wang Q, Guan Y, Wang J (2012) A chance-constrained two-stage stochastic pro-
              gram for unit commitment with uncertain wind power output. IEEE Trans Power Syst
              27(1):206–215
[WKK10]   Wang J, Kennedy S, Kirtley J (2010) A new wholesale bidding mechanism for
              enhanced demand response in smart grids. In: Innovative smart grid technologies,
              pp 1–8
[WSF10]   Wu L, Shahidehpour M, Fu Y (2010) Security-constrained generation and transmission
              outage scheduling with uncertainties. IEEE Trans Power Syst 25(3):1674–1685
[WWG13a]   Wang Q, Wang J, Guan Y (2013) Stochastic unit commitment with uncertain demand
              response. IEEE Trans Power Syst 28(1):562–563
[WZLZ14]   Wu Z, Zhou S, Li J, Zhang X-P (2014) Real-time scheduling of residential appliances
              via conditional risk-at-value. IEEE Trans Smart Grid 5(3):1282–1291
[ZP10]   Zheng QP, Pardalos PM (2010) Stochastic and risk management models and solution
              algorithm for natural gas transmission network expansion and lng terminal location
              planning. J Optim Theory Appl 147:337–357
[ZSS15]   Zheng QP, Shen S, Shi Y (2015) Loss-constrained minimum cost flow under arc failure
              uncertainty with applications in risk-aware kidney exchange. IIE Trans 47(9):961–977
[ZWPG13]   Zheng QP, Wang J, Pardalos PM, Guan Y (2013) A decomposition approach to the
              two-stage stochastic unit commitment problem. Ann Oper Res 210:387–410
[ZWWG13]   Zhao C, Wang J, Watson JP, Guan Y (2013) Multi-stage robust unit commitment con-
              sidering wind and demand response uncertainties. IEEE Trans Power Syst 28(3):2708–
              2717
[ZZ12]   Zhao L, Zeng B (2012) Robust unit commitment problem with demand response and
              wind energy. In: IEEE power and energy society general meeting, pp 1–8

Appendix A
Nomenclature

UC	Unit Commitment
SCUC	Security-Constrained Unit Commitment
SCRA	Security-Constrained Reliability Assessment
SUC	Stochastic Unit Commitment
ED	Economic Dispatch
DR	Demand Response
ES	Energy Storage
ISO	Independent System Operator
RTO	Regional Transmission Organization
BD	Benders' Decomposition
LR	Lagrangian Relaxation
RMP	Relaxed Master Problem
SP	Subproblem
LB	Lower Bound
UB	Upper Bound
DAM	Day-Ahead Market
RTM	Real-Time Market
RTC	Real-Time Commitment
RTD	Real-Time Dispatch
LMP	Locational Marginal Price
RAA	Reserve Adequacy Assessment

A.1 Sets and Indices

A Set of power transmission lines

G Set of all thermal generators

G_i Set of thermal generators at bus i

N Set of locations (buses)

T Length of power generation planning horizon, i.e. 24 h

© The Author(s) 2017
Y. Huang et al., *Electrical Power Unit Commitment*,
SpringerBriefs in Energy, DOI 10.1007/978-1-4939-6768-1

g Indices of thermal generators
i, j Indices of buses
t Time period
$\varXi$ The set of all simulated scenarios
ξ Indices of scenarios

A.2 Parameters

SU_{gt} startup cost of generator g in period t
SD_{gt} shutdown cost of generator g in period t
$Prob^{\xi}$ probability of scenario ξ
L_g minimum ON time of generator g
l_g minimum OFF time of generator g
P_g^{max} maximum power generation capacity for generator g
P_g^{min} minimum power generation capacity for generator g
RU_g ramping up limit of generator g
RD_g ramping down limit of generator g
RS_{it} spinning reserve requirement at bus i in period t
S_g^{max} maximum spinning reserve of generator g
R_{it}^{ξ} renewable energy output at bus i in period t of scenario ξ
D_{it} forecasted hourly demand at bus i in period t
E_{it}^{ξ} price elasticity coefficient at bus i in period t of scenario ξ
ρ_i energy storage efficiency at bus i
B_{ijt} susceptance in branch i ? j in period t ;
θ confidence level in risk constraints
β_{it}^{ξ} voltage angle at bus i
ϕ maximum load-shedding loss allowance
α, γ electricity price velocity indicators
κ_i power storage capacity at bus i

A.3 Variables

u_{gt} commitment decision of generator g at period t
v_{gt} startup action of generator g at period t
w_{gt} shutdown action of generator g at period t
p_{gt}^{ξ} power generation of generator g in period t of scenario ξ
s_{gt}^{ξ} spinning reserve of generator g in period t of scenario ξ
f_{ijt}^{ξ} power flow from bus i to bus j in period t of scenario ξ
q_{it}^{ξ} recommended electricity price at bus i in period t of scenario ξ

r_{it}^{ξ} remaining power in power bank at bus i in period t of scenario ξ

v_{it}^{ξ} power charging amount in power bank at bus i in period t of scenario ξ

x_{it}^{ξ} power dispatch amount in power bank at bus i in period t of scenario ξ

y_{it}^{ξ} shifted demand at bus i in period t of scenario ξ

η_{t} Value-at-Risk in period t (VaR)

ζ_{t}^{ξ} load-shedding loss over VaR in period t of scenario ξ

Appendix B
Renewable Energy Scenario Generation

Here, we introduce a simple method for scenario generation in C++. Scenario generation is initially to generate a sequence of random numbers following a specific distribution, such as Normal distribution or Weibull distribution, and then randomly select a proportion of scenarios to construct a scenario set.

We applied this scenario generation method to support the problem modelling in Chap. 3. Since the wind energy output is assumed to follow a normal distribution, which is described by the probability density function:

$$p(x|\mu, \sigma) = \frac{1}{\sigma\sqrt{2\pi}} \cdot e^{-\frac{(x-\mu)^2}{2\sigma^2}} \tag{B.1}$$

The distribution parameters thus are input including mean (μ) and stand deviation (σ). The procedure of random number generation has two steps [Cpl11]:

- a generator produces sequences of uniformly distributed numbers;
- a distribution transforms above numbers into sequences of numbers with a specific distribution.

Let $x \sim N(0, 100)$, the C++ codes for scalable scenario generation are shown as follow.

```
typedef  std::tr1::ranlux64_base_01            ENG;
typedef  std::tr1::normal_distribution <double> DISTA;
typedef  std::tr1::variate_generator <ENG,DISTA> GENA;

double x;
ENG eng;
    eng.seed((unsigned int)time(NULL));

for(i=0;i<numscn;i++)
            for(k=0;k<numbus;k++)
                    DISTA  dist(0,10);
                    GENA   gen(eng,dist);
                    x = 0; dist.reset();
                    x = gen();
```

© The Author(s) 2017
Y. Huang et al., *Electrical Power Unit Commitment*,
SpringerBriefs in Energy, DOI 10.1007/978-1-4939-6768-1

Table B.1 10 scenarios of wind energy outputs

Mean	S1	S2	S3	S4	S5	S6	S7	S8	S9	S10
45.0	48.4	59.1	49.3	43.4	46.5	34.1	43.5	41.6	52.4	42.1
51.0	54.4	65.1	55.3	49.4	52.5	40.1	49.5	47.6	58.4	48.1
58.0	61.4	72.1	62.3	56.4	59.5	47.1	56.5	54.6	65.4	55.1
36.0	39.4	50.1	40.3	34.4	37.5	25.1	34.5	32.6	43.4	33.1
39.0	42.4	53.1	43.3	37.4	40.5	28.1	37.5	35.6	46.4	36.1
34.0	37.4	48.1	38.3	32.4	35.5	23.1	32.5	30.6	41.4	31.1
43.0	46.4	57.1	47.3	41.4	44.5	32.1	41.5	39.6	50.4	40.1
41.0	44.4	55.1	45.3	39.4	42.5	30.1	39.5	37.6	48.4	38.1
33.0	36.4	47.1	37.3	31.4	34.5	22.1	31.5	29.6	40.4	30.1
31.0	34.4	45.1	35.3	29.4	32.5	20.1	29.5	27.6	38.4	28.1
28.0	31.4	42.1	32.3	26.4	29.5	17.1	26.5	24.6	35.4	25.1
28.0	31.4	42.1	32.3	26.4	29.5	17.1	26.5	24.6	35.4	25.1
30.0	33.4	44.1	34.3	28.4	31.5	19.1	28.5	26.6	37.4	27.1
31.0	34.4	45.1	35.3	29.4	32.5	20.1	29.5	27.6	38.4	28.1
33.0	36.4	47.1	37.3	31.4	34.5	22.1	31.5	29.6	40.4	30.1
24.0	27.4	38.1	28.3	22.4	25.5	13.1	22.5	20.6	31.4	21.1
20.0	23.4	34.1	24.3	18.4	21.5	9.1	18.5	16.6	27.4	17.1
31.0	34.4	45.1	35.3	29.4	32.5	20.1	29.5	27.6	38.4	28.1
33.0	36.4	47.1	37.3	31.4	34.5	22.1	31.5	29.6	40.4	30.1
38.0	41.4	52.1	42.3	36.4	39.5	27.1	36.5	34.6	45.4	35.1
41.0	44.4	55.1	45.3	39.4	42.5	30.1	39.5	37.6	48.4	38.1
43.0	46.4	57.1	47.3	41.4	44.5	32.1	41.5	39.6	50.4	40.1
44.0	47.4	58.1	48.3	42.4	45.5	33.1	42.5	40.6	51.4	41.1
41.0	44.4	55.1	45.3	39.4	42.5	30.1	39.5	37.6	48.4	38.1

```
if (k==0)
        for(j=0;j<numhr;j++)
                wind[j][k]= mean[j]+x;
else
        for(j=0;j<numhr;j++)
                wind[j][k] = 0;

dist.reset();
    end
end
```

Given the mean of wind energy output for each hour, we generate a hundred of scenarios and randomly select 10 scenarios, shown in Table B.1.

Reference

[Cpl11] Cplusplus.com.   Random - c++   reference   (2011).
         http://www.cplusplus.com/reference/random/